ESSAYS ON WELLNESS

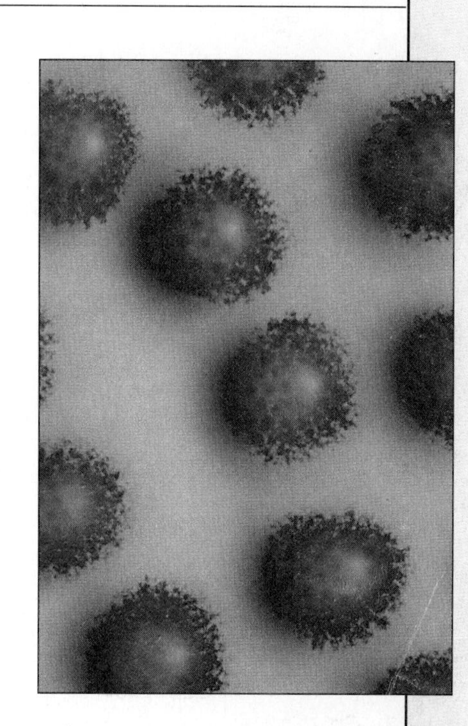

Barbara A. Brehm
Smith College

HarperCollins*CollegePublishers*

Acquisitions Editor: Bonnie Roesch
Developmental Editor: Meryl R. G. Muskin
Project Editor: Thomas R. Farrell
Art Director: Teresa J. Delgado
Text Design: Baseline Graphics
Cover Design: Delgado Design, Inc.
Cover Photo: Gary Nolton ©1991
Electronic Page Makeup: Circa 86, Inc.
Photo Researcher: Mira Schachne
Production Administrator: Jeffrey Taub
Printer and Binder: Von Hoffman Press, Inc.
Cover Printer: The Lehigh Press, Inc.

ISBN 0-06-501549-5

93 94 95 9 8 7 6 5 4 3 2

Contents

Preface

The wellness philosophy is based on the idea that the life-style choices we make throughout the years have an important impact on our mental and physical well-being. *Wellness* means doing what you can to maximize your personal potential for optimal well-being and to construct a meaningful and rewarding life. It acknowledges the important link between health and behavior. While disease prevention is an important goal of the wellness life-style, *wellness* implies more than a state of good health.

Wellness means taking responsibility for our health, preventing accidents and illness, and knowing when to consult health-care professionals. Wellness encourages consumer awareness and promotes the establishment of social systems and environments conducive to health-promoting behavior.

Why Wellness?

This book is a collection of 29 essays on a variety of wellness topics. Topics range from the traditional personal health areas of exercise, stress management, and nutrition to medical applications such as infertility, arthritis, and the safety of blood transfusions. Each essay provides a brief summary of our current understanding of a particular topic, synthesizing recent medical and scientific research. The first essay introduces the concept of wellness in the context of human biology and health and provides a logical beginning for the readings that follow. Other than that, essays are independent of each other and may be read in any order.

These essays contain applications to many fields and offer a multidisciplinary perspective. Some instructors may wish to use *Essays on Wellness* as a supplementary text for courses in health, human biology, and anatomy and physiology. *Essays on Wellness* adds interest to science and health courses. It helps the instructor answer that common classroom question: *So what?* Personally relevant information provides students with an anchor for broader conceptual knowledge and motivates them to achieve a deeper understanding of the subject at hand.

Barbara A. Brehm

ESSAY 1

LIFE-STYLE, HEALTH, AND HOMEOSTASIS

> What a piece of work is man!
> How noble in reason, how infinite in faculties, in form and moving how express and admirable, in action how like an angel, in apprehension how like a god! The beauty of the world, the paragon of animals!
>
> *William Shakespeare*

Good health: homeostasis is the basis

The miracle of life seems no less wondrous when one begins to explore the intricacies of human biology and health. Even as we approach today's limits of scientific understanding, we cannot help but marvel at the beauty of creation.

One of the most interesting and important processes of life is the ability of an organism to maintain homeostasis. Homeostasis is a condition in which the internal environment of an organism remains within certain physiological limits in terms of temperature, chemical composition, and pressure. When these limits are exceeded, the organism compensates in some way to return to homeostasis. For example, if blood sugar falls too low, certain hormones are secreted that facilitate the release of sugar from glycogen stores in the muscles and liver, bringing the blood sugar up into normal range. Homeostasis is regulated by the nervous and endocrine systems, which allow all of the body's systems to communicate with one another. Homeostasis is the basis of good health. An understanding of homeostasis is helpful for understanding the nature of health and disease.

One might view the human body as a large orchestra, and its life processes as the music the orchestra produces. The nervous and endocrine systems act together as the conductor to direct the various orchestral sections to keep all parts blended into a harmonious balance. The quality of the music suffers if even one of the instruments is off-key or plays at the wrong time. The study of anatomy and physiology is like learning about the instruments in the orchestra and how they all play together.

The term *homeostasis* does not imply that the body stays the same, that it exists in a static state. Rather, homeostasis is a dynamic balance, a balance among the components of the body and between the body and its environment. Internal and external events demand continuous physiological adjustments. The body is in a continuous state of flux.

Health behavior

The body's ability to maintain homeostasis gives it tremendous healing power and a remarkable resistance to abuse. But, for most people, lifelong good health is not something that just happens. Two important factors in this balance we call health are the environment and behavior of the organism in question. Our homeostasis is affected by the air we breathe, the food we eat, and even the thoughts we think.

Disease can result from a disruption of homeostasis, which may be brought on by one's behavior and interaction with the environment. The way we live our lives can either support or interfere with the body's ability to maintain homeostasis. Many diseases are the result of years of poor health behavior. An obvious example is smoking. Smoking tobacco exposes sensitive lung tissue to a multitude of chemicals that cause cancer and damage the lung's ability to repair itself. Since lung cancer is difficult to treat and very rarely cured, it is much wiser to quit smoking (or never start) than to hope the doctor can fix you up once you are diagnosed with this disease.

The health-care system's power is limited when it comes to treating many life-style diseases such as lung cancer, emphysema, obesity, and artery disease. Health-care professionals face frustration daily in their attempts to treat patients who resist treatment or are unable to change the life-style habits necessary to promote effective healing.

Coronary artery disease

Heart disease provides one of the best illustrations of the connection between life-style and disease. Blood flow to the heart is threatened when the arteries of the heart become clogged with plaque, a process called *atherosclerosis.* Heart attacks and their complications are still the leading cause of premature death in Canada and the United States.

Many factors contribute to the process of atherosclerosis, and some of them are under an individual's control. The four most important controllable factors are physical activity level, smoking, blood cholesterol levels, and blood pressure. People can substantially reduce their risk of heart disease by not smoking and by controlling blood cholesterol and blood pressure levels through appropriate diet, exercise, weight control, and, if necessary, medication. Statistics show that we have improved our control of these variables. Fewer North Americans are smoking. We're eating less fat, exercising more, and keeping an eye on cholesterol and blood pressure. Mortality rates for heart disease have declined dramatically over the past 20 years, and experts believe this decline is due largely to improvements in life-style behaviors. There is still plenty of room for improvement, but we've shown that life-style can make a difference.

Beyond health

For many years, the word *health* was used to mean simply an absence of disease. If you were free of medical symptoms requiring a physician's intervention, you were healthy. Many people were uncomfortable with this notion, and felt that there was more to health than simply not being sick. The term *wellness* came into use to indicate that the health continuum (illustrated in the accompanying figure) represents more than the presence or absence of disease. And just as life-style can make us sick, it can also help make us well.

Cardiologist George Sheehan, a devout believer in the value of health behavior, once said that when he thought about life expectancy, he preferred to think about what he expected from life. Health is more than the absence of disease. The word *wellness* is often used to mean optimal health and living the sort of life-style that engenders it.

Wellness means taking responsibility for one's health, preventing accidents and illness, knowing when to consult a health-care professional, and working with health-care providers when necessary. Wellness encourages consumer awareness and promotes the establishment of social systems and environments conducive to health-promoting behavior. Paradoxically, disease and disability do not prevent a wellness life-style, for wellness simply means maximizing one's potential for physical, psychological, and social well-being.

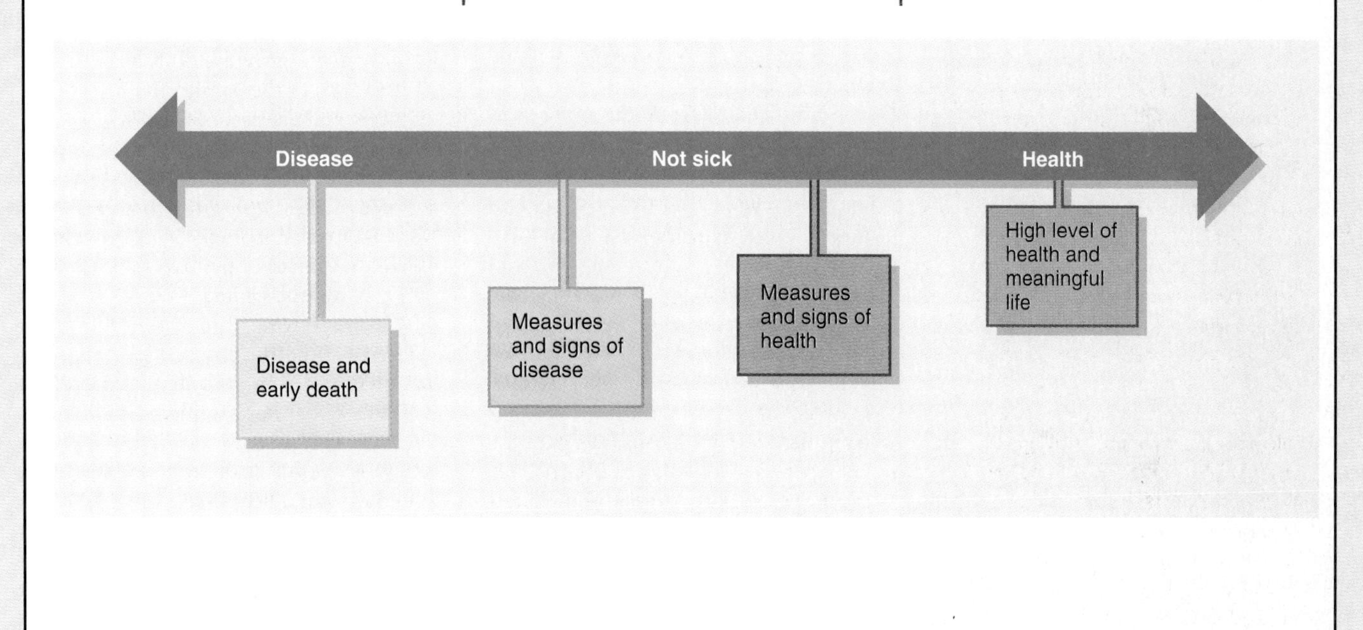

ESSAY

2 VEGETARIAN DIETS: ALPHABET SOUP

Health benefits of a vegetarian diet

Research continues to show that vegetarians are generally healthier than their meat-eating friends. Not only do they, as a group, have less artery disease, they also have lower rates of high blood pressure, diabetes, and certain types of cancer; a healthier digestive system; and lower cholesterol and body fat levels. Scientists attribute much of this difference to the vegetarian's low-fat, high-fiber diet. A well-planned vegetarian diet is also high in vitamins and minerals, and low in cholesterol.

So, while 30 years ago most North Americans thought vegetarianism was only for "health nuts," the current quest for a low-fat life-style has promoted widespread interest in the vegetarian style of eating among all kinds of people. Some become lacto-ovo vegetarians, people who do not eat meat but continue to eat eggs and dairy products. Others become vegans, people who eat no animal products at all. Many people continue to eat meat but are choosing to eat less of it and to have several meatless meals each week. People new to vegetarian meal planning sometimes wonder if there is anything special they should know to be sure they get all the nutrients they used to get from meat.

Protein quality

Since meat is a concentrated source of protein, protein is often the new vegetarian's first concern. Dietary protein provides the amino acids used to build proteins in the body. Our bodies use at least 20 different amino acids to manufacture the wide variety of proteins needed. Given an adequate protein intake, we can make 10 of the amino acids we need. The other 10 are called *essential amino acids*, which means we must obtain them from our diet.

Animal products, including meat, eggs, and dairy products, are complete proteins; that means they contain all 10 essential amino acids in the relative proportions needed by the human body. Plant sources, on the other hand, are often low on 1 or 2 essential amino acids. You can see why vegetarians have been concerned if you consider the analogy of proteins to words made from amino acid letters. Even if you had plenty of *Y*s and *S*s, you still couldn't spell *YES* if you were missing the letter *E*.

Is it difficult to get all the amino acid letters without eating meat? Fortunately, it is easy to obtain enough of the right kinds of amino acids from plant sources as long as a wide variety are consumed. For example, the amino acids lacking in rice are found abundantly in beans. Rice and beans are an example of complementary proteins. Consume the two together, or separately during the same day, and you will get all the amino acids necessary for good health.

To ensure an adequate intake of all 10 essential amino acids from plant sources, it is helpful to think of three groups of protein foods: legumes, grains, and nuts/seeds. While foods in each group provide incomplete protein, the missing amino acids will be provided if a food from one of the other groups or dairy products or eggs is eaten as well. Rice and bean casseroles, peanut butter and whole-wheat bread, and chili and cornbread are examples of food combinations that provide high-quality protein.

The more restricted a person's diet, the more careful he or she needs to be about menu planning. A person who simply forgoes meat two or three days a week and has a generally healthful diet need not be more concerned about nutritional deficiencies than any other person. Lacto-ovo vegetarians can obtain all the nutrients found in a meat-eater's diet. They must be careful, however, not to rely solely on eggs and dairy products for their protein, or they may consume too much fat and cholesterol.

Other nutritional concerns

Vegans must be especially careful to include a wide variety of foods to be sure they get all the amino acids they need, as well as the vitamins and minerals others get from eggs, dairy products, and meat. Population groups with special nutritional needs and considerations, such as pregnant and lactating women, infants, and children, are especially vulnerable to possible nutritional deficiencies imposed by a restricted diet. The following nutrients are common in animal products, and more difficult to obtain on a vegan diet. Epidemiological studies show that populations with marginal intakes of these nutrients seem to have adapted to these low intakes, and rarely show signs of deficiencies. Nutritionists are unsure how long it takes for such adaptation to occur, and generally recommend that new vegetarians, especially vegans, keep an eye on these nutrients:

Vitamin B_{12}: Found only in animal foods and in nutritional supplements. Vegans can obtain vitamin B_{12} from nutritional yeast, or from vitamin B_{12}-fortified products such as soy milk.

Iron: Vitamin C increases iron absorption from plant sources, such as legumes, grains, and vegetables. People with anemia (most commonly, premenopausal women and women during pregnancy) may need a supplement.

Zinc: Found to some extent in grains and legumes.

Vitamin D: Plant sources do not

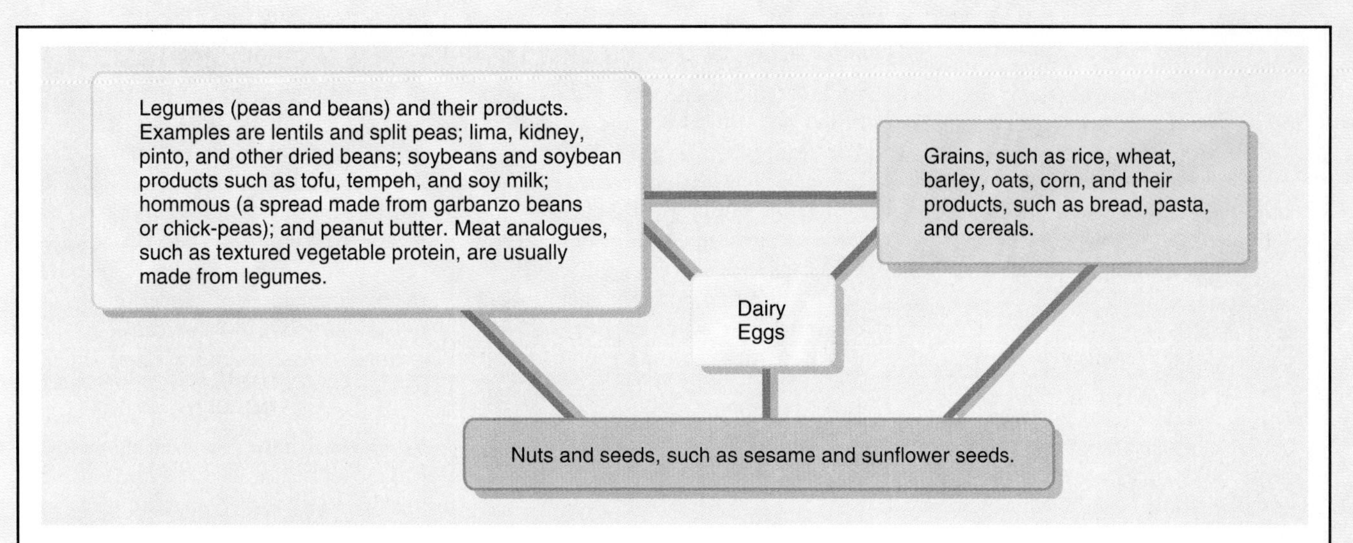

provide vitamin D, although regular exposure to the sun will prevent a deficiency. Vitamin D supplements may be necessary for vegans who live in cloudy climates or don't get outdoors much.

Riboflavin: Present in dark green vegetables, so vegans who consume good amounts of these foods are able to meet their needs for this B vitamin.

Calcium: Found in dark green vegetables, sesame seeds, legumes, and grains. Since daily requirements for this mineral are fairly high, it is often difficult to meet calcium needs through plant sources alone. For this reason, many vegans rely on calcium-fortified soy milk. An adequate calcium intake is especially crucial for children and pregnant and lactating women.

It's important to note that simply eliminating meat from one's diet does not automatically make it a healthful one. There are plenty of fatty foods free of animal products, so you can give up hamburgers and still get an excess of fat from french-fried potatoes, milkshakes, and grilled cheese sandwiches.

ESSAY 3

THE CANCER-PREVENTION LIFE-STYLE

While the process involved in the development of cancer is only partially understood, it is known that cancer develops through contact with one or more carcinogens that cause the genetic changes that allow cells to grow out of control.

It is believed that there are two types of carcinogens involved in causing most cancers. Initiators start the cellular damage that can lead to cancer, and promoters allow genetically damaged cells to proliferate at a greater rate than the normal cells. For example, alcohol promotes cancer of the mouth and esophagus when combined with an initiator such as tobacco.

Scientists at the National Cancer Institute estimate that about 80% of all cancer cases are related to life-style. The good news is that some life-style factors actually prevent the development of cancer. Many cancers are preventable by avoiding carcinogens and following recommendations for a cancer-prevention life-style.

Tobacco use

Former U.S. Surgeon General C. Everett Koop called tobacco use America's single most preventable cause of death. Tobacco use contributes to the three leading causes of death in North America: artery disease, cancer, and stroke.

Cigarette smoking causes 30% of all cancer deaths. About 90% of lung cancer patients are smokers. Cigarette smoking also causes cancers of the larynx, esophagus, pancreas, bladder, kidney, and mouth. Low-tar and low-nicotine cigarettes are no solution, because they actually increase the smoker's risk of cancers of the mouth and throat. Pipe smoking increases the smoker's risk of cancers of the mouth, tongue, and throat, and chewing tobacco increases the user's risk of cancer of the mouth.

Tobacco smoke contains hundreds of damaging chemicals, and includes both cancer initiators and promoters. For example, the phenols found in tobacco tar greatly increase the carcinogenic potency of benzopyrene, another substance found in cigarette tar. (Cigarette tar refers collectively to several hundred different chemicals in cigarettes that, when condensed, form a brown, sticky substance.)

Alcohol

Heavy alcohol use (more than two drinks a day) is associated with cancers of the mouth, throat, esophagus, and liver. (One drink contains about 0.6 oz of alcohol, and is comparable to 1.5 oz of liquor, a 5-oz glass of wine, or a 12-oz glass of beer.) People who drink and smoke have a much greater risk of getting cancers of the mouth and esophagus. Alcohol may also contribute to the development of breast cancer.

Sexual contact

Viruses may act as carcinogens. Many of the viruses associated with an increased risk of cancer [such as human papilloma virus (HPV)] are spread through sexual contact. Sexually active people should follow safer-sex practices, such as using condoms during sexual contact, to reduce the likelihood of getting these viruses.

Diet

Several dietary factors are associated with a person's risk of developing cancer. Carcinogens are found in foods that are smoked, cured, and pickled. Consumption of pesticide residues on fruits and vegetables and in meats may increase an individual's cancer risk.

Fat intake may be the major and most controllable dietary carcinogen in North America. Fat appears to act as a cancer promotor, especially in cancers of the breast and colon. These are the leading causes of cancer deaths in nonsmoking North Americans.

Some dietary chemicals may block cancer promotion: retinoids, vitamins C and E, and the mineral selenium. Researchers have theorized that some carcinogens cause cancer by producing highly reactive oxygen atoms called *free radicals,* which may cause damage to cell components, such as DNA. Retinoids, vitamins C and E, and selenium appear to act as antioxidants; that is, they neutralize these reactive chemicals and thus block their carcinogenic effect.

Retinoids can be converted to vitamin A in the body. While too much vitamin A is toxic, retinoids in quantity appear to be safe. Good sources of retinoids include dark green vegetables such as spinach and broccoli, and yellow fruits and vegetables such as cantaloupe, apricots, carrots, and yams.

Vitamin C is found in many fruits and vegetables such as citrus fruits, strawberries, and potatoes. Vitamin E, a normal component of cell membranes, is found in vegetable oils, and selenium is found in seafood, whole grains, and organ meats.

Studies on both humans and animals suggest that fiber may help prevent cancers of the colon and rectum. Fiber is plant material that people can't digest. These materials may be divided into two categories: water soluble and water insoluble. It is the insoluble type, which comes from the structural components of plants, that is linked with a healthy digestive tract. Insoluble fiber is found in whole-grain cereals, wheat bran, breads, and many vegetables.

A family of plants known as cruciferous vegetables also contain antioxidants and have received some attention as possible cancer preventers. These vegetables include broccoli, brussels sprouts, cabbage, and cauliflower.

Weight control and physical activity

Life-insurance data indicate that people who are more than 20% above their recommended weight have a higher than average risk of many types of cancer. Cancer risk increases with the amount of extra weight.

Several studies have found an association between low levels of occupational and/or recreational physical activity and risk of colon cancer. Sedentary participants in one study had two to three times the risk of colon cancer as more active participants. The intensity of the activity does not appear to be an important variable. People who performed even mild physical activity, such as walking, had a lower cancer risk than sedentary folks. Preliminary evidence also suggests that active women may have fewer cancers of the breast and reproductive system.

Exposure to carcinogens

Carcinogens are present in many industrial and household products. Regulatory agencies, industries, and organized labor have developed procedures for reducing hazardous exposure to carcinogens in the workplace. Household and garden products should always be used as directed to reduce exposure to fumes and contact with skin.

Ultraviolet light is a potent carcinogen, especially for light-skinned people. Protective clothing and sunscreens can prevent much of the carcinogenic effect of sunlight.

Personality and stress

Some studies have found that cancer is more common among people who keep their anger and other emotions bottled up and who cope poorly with internal turmoil and stress. The association between stress and disease is explored in greater detail in Essay 22.

ESSAY

FAT TISSUE ISSUES

4

Adipose tissue: No other body tissue has come close to achieving such popular attention, fame, and notoriety. It has been featured in almost every major newspaper and magazine, and provides a recurring theme in some magazines as a focus of fashion and fascination. No other body tissue has inspired the creation of more products claiming to compress, massage, shrink, and melt it.

Adipose tissue has even sparked political and philosophical controversy. Fat people have rallied to expose the prejudice and discrimination they face in our culture. Feminists argue that women are judged too frequently by how much fat they have and by its anatomical location. Physiological gender differences (even lean females have about 50% more fat than males) in body composition often conflict with fashion ideals, leading young girls to develop negative body images, low self-esteem, and eating disorders. This has led some women to say that fat is a feminist issue (since, as a physiology professor once pointed out, fat is a feminine tissue).

Fashion aside, fat performs several vital physiological functions. The fat incorporated into such organs and tissues as mammary glands, nerves, brain, and lungs and the fat in adipose tissue that cushions and supports vital organs is known as *essential fat*. Storage fat is considered nonessential, although it has the important function of storing energy-rich molecules.

Too much of a good thing

Obesity refers to excess body fat, and is usually defined as being 20% or more overweight. About 30% of the people in the U.S. fall into this category.

Life-insurance data have shown a negative association between excessive fatness and mortality, and the desirable weight-for-height charts have been derived from these data. People falling into the weight ranges for a given gender and height on these charts generally have a longer life expectancy.

Obesity alone may not pose a risk of premature death, but the frequency of several serious health problems does increase with excess weight. Obesity increases the risk of hypertension, type II diabetes mellitus, and hyperlipidemia (high serum cholesterol and/or triglycerides), all three of which increase a person's risk of heart disease and stroke. Gallstones, gall bladder disease, cirrhosis of the liver, kidney disease, and some cancers are more common in obese individuals. In addition, extra fat results in increased stress on the weight-bearing joints, with higher rates of osteoarthritis, back pain, and other orthopedic complications.

How much is too much?

Height–weight charts give a useful range for healthful body weights, but weight alone can be misleading. While the scale does not lie, it doesn't tell the whole truth either. Weight is not the same thing as fat. For most sedentary people, a weight gain does mean an increase in body fat. But it is also possible to get fatter over the years and see no change in weight. Sedentary adults may weigh the "correct" amount, but be overly fat because their muscles have atrophied from disuse. Likewise, people who embark on a vigorous exercise program may lose fat yet see no change or even an increase in body weight. Bodybuilders and other strength athletes may produce a hefty number on the scale, and weigh more than they "should" for their height, yet be relatively lean.

To get around the limitations of the height–weight tables, various methods for assessing fatness, or body composition, have been devised, based on the physiological properties of adipose tissue. These methods divide body composition into two components: fat and everything else or fat-free mass. Body composition is often expressed as the percentage of the body's volume that is fat. The desirable body fat range for men is 12 to 20%, with athletes as low as 5%. Women should be about 20 to 30% fat, although female athletes may be 16% or less.

Most people know whether they have too much adipose tissue without a body composition assessment. But such assessments can help an obese person set a realistic weight (fat) loss goal and help monitor body composition change in obesity research. Many fitness centers offer some sort of body composition assessment to help motivate clients to exercise.

Haunch versus paunch

All fat stores are not created equal. Research has found that extra fat on the torso is associated with a greater health risk than extra fat on the hips and thighs. Central adiposity is more common among males, although it occurs in women as well, and may be one of the reasons that males have a higher risk of cardiovascular disease.

The health risks associated with having an "apple" versus a "pear" shape include insulin resistance and type II diabetes mellitus, high blood cholesterol, and high blood pressure, all of which increase a person's risk of heart disease.

Body shape, or fat pattern distribution, can be assessed very simply by comparing waist and hip girths. A waist–hip ratio (WHR) greater than 1.0 for men or 0.8 for women is considered to put an individual at increased risk for the above disorders.

Adipocytes in different regions appear to be metabolically different. Fat cells in the abdominal region release directly into the vein that goes to the liver, the site of a great deal of fat, protein, and carbohydrate metabolism, and the place where cholesterol-containing lipoproteins are produced. Abdominal fat cells may thus have a stronger impact on fat and cholesterol metabolism than cells in more peripheral locations. Studies have also shown that abdominal adipocytes also respond more readily to several hormones, including adrenalin.

Fat pattern distribution appears to be genetically determined, and is not influenced by diet or exercise. If an obese person loses weight, fat is generally lost from all areas of the body. As one researcher put it, "An individual continues to resemble himself in relative fat pattern, despite weight loss."

Many people persist in performing calisthenics in a misguided attempt to influence fat deposits in specific body areas. "Spot reduction" does not occur, however. Sit-ups will strengthen abdominal muscles, but will not reduce the amount of adipose tissue stored in that area.

ESSAY

MORE THAN SKIN DEEP

5

The skin reflects a person's general health and well-being. What's good for your health is often good for your skin. And the health benefits of skin care are more than skin deep, since what affects the skin can affect other parts of the body and a person's self-image as well. Our skin is the most visible organ of our bodies, so skin disorders can have a large emotional impact. A healthy life-style can help keep skin healthy, reduce the severity of many dermatological disorders, and slow skin aging. When skin disorders develop, appropriate medical advice must be accompanied by educated, daily self-care.

Nutrition

The best nutritional recommendation for healthy skin is not magic, just plain good sense. It's the same phrase you've heard a hundred times before: Eat a well-balanced diet. But while true vitamin and mineral deficiencies are associated with skin disorders, getting more than the recommended daily allowance of vitamins and minerals does not make above-average skin. Supplements will improve the skin only if they are correcting an existing deficiency. In fact, too much of some vitamins can actually cause skin problems. For example, too much vitamin A is toxic and can make the skin rough and dry. Niacin can cause flushing and itching. And vitamin E applied topically can cause acne and allergic reactions in some people.

Food sensitivity

Some skin disorders are linked to food sensitivities. One person in five experiences hives at some time, which may be brought on by many things, including certain foods. Eggs, nuts, beans, chocolate, strawberries, tomatoes, citrus fruit, seafood, corn, and pork are the most common problematic foods. Several food additives may also cause hives. Tartrazine (yellow dye No. 5), the preservative sodium benzoate, and sulfites are common triggers. Salicylates also cause hives in some people. Salicylates are found in all aspirin-containing medications. They also occur naturally in almonds, apples, peaches, potatoes, and other foods.

Some evidence suggests that some foods exacerbate atopic eczema, or dermatitis. In one study of 36 children with eczema, 12 experienced significant improvement when eggs and cow's milk were eliminated from their diets.

Diet and acne

It is a commonly held belief that foods such as chocolate, cola drinks, nuts, and dairy products can make acne worse. Most research to date has failed

to establish a clear link between diet and acne. However, some acne patients and their physicians claim to have isolated foods that do affect acne severity, so it remains possible that diet, especially one high in fat and refined carbohydrates, may be a contributing factor in some cases.

Stress

Why is it that students (and teachers) look so much better in September than in December? Stress may be one of the factors responsible for this difference. Stress tends to worsen many preexisting skin problems, especially herpes, acne, eczema, hives, psoriasis, and warts. Stress-management techniques, including hypnosis, biofeedback, and other relaxation exercises, have been shown to improve many skin problems. These techniques may work by changing hormone levels and nervous system activity.

Noted scientist Lewis Thomas has written that it's "one of the great mystifications of science: warts can be ordered off the skin by hypnotic suggestion." The skin responds to a person's beliefs. In one study, subjects were told their skin was being exposed to poison ivy, and many broke out even though the poison ivy was only in their imaginations.

A person's behavior often makes a skin problem worse. Squeezing pimples and scratching irritated skin are two common examples. Relaxation techniques can be used to help lessen the picking and scratching, and also to decrease the itching associated with conditions such as eczema.

Smoking

People who smoke cigarettes usually show premature aging of the skin, especially facial wrinkles. Longtime smokers tend to look about 10 years older than nonsmokers of the same age. This effect may be related to some of the chemicals in cigarette smoke and to the decrease in peripheral blood flow caused by nicotine, a potent vasoconstrictor.

Exercise

During exercise, the body shunts blood to the skin to help release excess heat produced by the contracting muscles. This increased blood flow provides the skin with nutrients and gets rid of wastes. One study found that regular exercisers had thicker skin than sedentary individuals. Thicker skin ages more gracefully because it develops wrinkles later than thinner skin.

Sun protection

Protecting skin from the sun's damaging rays will help prevent premature aging and cancers of the skin, rates of which have been rising rapidly. In the U.S. over 500,000 people per year (that's one in seven Americans) will develop skin cancer during their lifetimes.

The sun's untraviolet rays are the source of skin damage. Until recently, scientists and consumers were concerned only about UV-B rays, since they are the ones that cause sunburn and skin cancer. UV-A rays cause tanning, and were once thought to be harmless. But UV-A rays actually penetrate the skin more deeply and can damage the skin's connective tissue, causing sagging and wrinkling of the skin. UV-A rays seem to increase the cancer-causing effects of UV-B rays.

Skin protection is the way to go outdoors. The most effective skin protection is some form of sun block. Tightly woven clothing (hold it up to a light and see how much shines through) helps keep the sun's rays from reaching the skin, and wide-brimmed hats provide some protection.

When a sun block is not practical, a sunscreen should be used. These do not shield the skin completely, but they do reduce the damaging effects of the ultraviolet rays. The degree of protection is indicated by a numerical rating, the sun protection factor (SPF), on sunscreen products. The SPF is the time a person can stay in the sun without burning when a product is used divided by the time a person could stay in the sun when the product is not used. For example, if your skin burns after 10 minutes in the sun without protection, but after 100 minutes with protection, the SPF of your sunscreen is 10. In other words, the sunscreen allows you to stay in the sun 10 times longer before burning. A sunscreen with SPF-15 is generally recommended, although people with light skin may need something stronger. Sunscreens and sun blocks should be used conscientiously whenever outdoors, even on cloudy days, since some radiation penetrates cloud cover.

ESSAY

6

HEALTHY BONES: MAKING AN IMPACT ON BONE DENSITY

Osteoporosis is a disorder characterized by decreased bone mass, owing to loss of bone mineral, and increased susceptibility to fractures. This disorder primarily affects middle-aged and elderly people, especially women. The female hormone estrogen protects against bone mineral loss during young adulthood. When estrogen levels decline after menopause (or because of a disruption in the menstrual cycle) bone mineral content declines as well.

Women of northern European descent are especially at risk for the development of osteoporosis, while African-Americans have a much lower risk. Small, thin females have higher rates of osteoporosis than large women. Small women have smaller bones to start with, and thin women have less adipose tissue, which is a great source of estrone, an estrogen that slows bone loss.

Though the loss of bone mineral may begin as early as the fourth decade of life, osteoporosis is often not diagnosed until so much bone mass has been lost that a person develops a fracture, which may have been caused by something no more traumatic than a sneeze. One of the most noticeable signs of osteoporosis is the rounded back posture known as "dowager's hump," so often depicted in the advertisements for calcium supplements. When the vertebrae lose bone mass, the spinal column loses strength, and individual vertebrae collapse because of what are known as crush fractures.

When these fractures accumulate in several vertebrae, the spine develops a hunchback that is often accompanied by back pain and loss of height.

While several factors affecting bone mineral density are outside your control, you can still make a significant impact on your lifelong bone density by cultivating a "high-density" lifestyle. When it comes to maintaining healthy bones, an ounce of prevention may be worth a pound of cure.

Calcium intake

Calcium intake has been the most publicized factor for promoting healthy bones, and the one with which Americans are probably most familiar. Several scientific and government groups have recommended increasing calcium intake to prevent osteoporosis, and the calcium supplement industry has not been slow to jump on the bandwagon.

A great deal of controversy surrounds the issue of calcium supplementation. Studies of whether calcium supplements help prevent or treat osteoporosis have yielded mixed results. Some studies have shown an increase in bone mineral density when calcium intake was increased. Other studies have failed to show a clear association between either childhood or adulthood calcium intake and the development of osteoporosis.

Nevertheless, nutritionists still agree that an adequate intake of dietary calcium is important for good health, which includes the growth and maintenance of bone tissue. Good sources of calcium include milk and milk products, dark green vegetables such as broccoli, sardines, and canned salmon (eat the bones too). Tofu and soy milk are often fortified with calcium for people who avoid dairy products.

Other foods and nutrients

Consumption of excess protein may contribute to loss of bone mineral. Doubling protein intake increases urinary calcium excretion 50%. This appears to be due to decreased recovery of calcium by the kidneys. Epidemiological studies have found lower rates of osteoporosis in British vegetarians and female Seventh-Day Adventists in the U.S. than in meat-eating individuals. These differences, however, could also be due to other factors, such as activity or heredity. But since most Americans consume more protein than they need, decreasing intake of concentrated protein sources such as meat and eggs (which are also high in cholesterol and fat) could be beneficial.

Some researchers have expressed concern that a very high-fiber diet may decrease calcium absorption from the intestine, since dietary fiber may bind with the calcium. Studies suggesting this concern have used fiber amounts much higher than those consumed by most people, so health professionals continue to recommend a moderate increase in fiber intake since such a diet is associated with many other health benefits. In countries where high-fiber diets are consumed, calcium balance and osteoporosis do not appear to be problems.

Vitamin D improves the efficiency of intestinal absorption of calcium. Enough vitamin D may not be synthesized, and it is difficult to obtain from natural dietary sources, so it is added to milk. Vitamin C, abundant in many fruits and vegetables, is essential for collagen production, and vitamin A, found in yellow and orange fruits and vegetables, also aids in bone development.

Several minerals in addition to calcium are important for bone development and maintenance. These include magnesium, found in seeds, grains, and green vegetables; manganese, found in blueberries, greens, and legumes; and boron, found in pears, apples, grapes, leafy vegetables, and legumes.

Frequent dieting can decrease bone density, as the body draws calcium from bone to make up for calcium missing from the diet. Therefore it is

likely that frequent periods of low caloric intake contribute to the development of osteoporosis.

Alcohol

The association between a high level of alcohol consumption and osteoporosis may be a consequence of poor nutrition, lower body weight, liver disease, or other illness, or it may be a result of a toxic effect of alcohol. A moderate intake of alcohol does not appear to cause bone loss.

Cigarette smoking

People who smoke cigarettes have a higher risk of osteoporosis than nonsmokers. This risk may be explained by the fact that smokers tend to be thinner, which is a risk factor. Female smokers also have lower estrogen levels and undergo menopause at an earlier age than nonsmokers, so they experience less of estrogen's protective effect.

Physical activity

Bone accommodates to the stresses imposed on it, so active people tend to have denser bones. Weight-bearing activity is strongly recommended to help increase bone mass during adolescence and young adulthood and to delay its loss thereafter. *Weight bearing* refers to activities in which the skeletal system must support body weight, for example, walking, hiking, and cross-country skiing as opposed to swimming. Progressive resistance training, the most common form of which is weight lifting, also appears to apply the mechanical stress required to increase bone mineral density.

Menstrual regularity

Menstrual irregularity in premenopausal women is often caused by eating disorders and/or excessive exercise, and signals low levels of estrogen. Several studies have documented osteoporosis in young women athletes who had stopped menstruating. Bone-density measurements have revealed that many of these young athletes had the bone density of women in their fifties. In some cases, the bone damage sustained was considered irreversible.

ESSAY

BACK BASICS

7

We usually take a healthy back for granted until something goes wrong. And statistics show that something often does: 8 out of 10 North Americans will experience back pain some time in their lives. Good back care means developing a healthy back by practicing good posture and body mechanics, and by maintaining adequate strength and flexibility of important postural muscles.

Poor posture and body mechanics can lead to chronic musculoskeletal problems. The way we sit, stand, drive, and perform common tasks such as lifting and carrying objects can lead to tight, painful muscles. These, in turn, can lead to inactivity, muscle inflexibility, weakness and imbalance, and a worsening of the poor posture and body mechanics that caused the problem in the first place.

The importance of posture

Most of us probably received our first posture advice at a very young age, perhaps when a well-meaning relative poked us in the back and admonished us to "stand up straight." Since grown-ups didn't know anything back then, many of us straightened up momentarily and then ignored the irrelevant remark.

Our primary opponent in the good-posture game is gravity. Gravity is always our partner, as we sit, stand, or move about. The muscles of our neck, for example, must exert enough force to lift and balance the head, which in an adult weighs about 10 to 12 pounds. Good posture results when our muscles maintain our skeletal frame in the best possible balance, a balance that places the least possible amount of stress on our joints. A person with good posture seems to make light of gravity.

What is good posture?

In general, our posture goal, whether sitting, standing, lifting, or dancing, is to maintain the normal shape of the spine. A healthy spine has a slight forward curve in the neck region (cervical curve), a backward curve in the upper back (thoracic curve), and another forward curve in the lower back (lumbar curve). The slight curve in the sacrum provides a fourth curve. Injury is more likely to result when these curves are either exaggerated or removed.

For good standing posture, the center of the head, shoulders, and hips should fall in a line when viewed from the side. The most common posture problems are rounded shoulders, a forward head (chin jutting forward), a protruding abdomen, and an exaggerated lower back curve (lordosis). It's not uncommon for all four to occur together.

When sitting, your back should have the same gentle curves as it does when standing. Slouching places pressure on the lower back that is 10 to 15 times greater than that placed on the joints when you're lying down. Use a chair that supports the curve in your lower back or add a small cushion. A back support cushion can also be helpful when driving. To prevent rounded shoulders when working at a computer or desk, put your work at eye level with some sort of support, so that you don't have to round your head forward. If your work requires several hours of sitting every day (students take note!), it's especially important to develop good posture habits.

Think about the best way to maintain the normal curves of your spine whenever you need to lift, push, pull, or carry objects. For example, when lifting a heavy object, like a child, from the floor, the lifter should try to lift "with the legs" by lowering himself or herself with bent knees, while keeping the back long, and never by lifting with a rounded back. When pushing or pulling something heavy, bend your knees, tighten your abdominal muscles to maintain correct spinal alignment, and then push or pull. Carry objects close to your body, and never with a rounded back.

Postural hazards

In adulthood, our daily life-style and posture habits have effects that accumulate with the years. People with sedentary occupations can prevent many back problems if they develop work stations that promote good alignment, practice habitual good posture and body mechanics, and maintain exercise programs to prevent the physical debilitation caused by inactivity.

Some occupations demand unusual physical stress. Dentists and hygienists, for example, may sit in twisted, rounded positions all day. Nurses, chiropractors, and physical therapists often do a lot of lifting and work in awkward positions. Musicians are notorious for having injuries related to long hours of practicing in awkward positions, such as while standing with a violin tucked under the chin. People who do physical work that involves lifting heavy loads, working with heavy machinery, and driving for long hours are also susceptible to back injury if they do not practice good posture and body mechanics.

Extra weight in the abdominal area puts a strain on the lower back. Both pregnancy and paunches can be a source of back problems. Many people find that getting rid of the extra weight (delivering the baby, or using up the excess fat) solves their back problem.

The aging process takes a heavy toll on posture, as bone mineral is lost from the vertebrae, and intervertebral discs "dry out," causing a loss of height and a more rounded thoracic curve (kyphosis). This in turn leads to muscle tightness and stiffness, with accompanying pain as the body adjusts to this unnatural alignment. Maintaining healthy bones (see Essay 6) is important for a healthy back as we age.

The importance of fitness

Research has shown that 80% of all patients who come to a physician with back complaints have no underlying organic disease. But back care experts have noted that they are usually deficient in the strength and flexibility of key postural muscles. These include the following:

1. **Abdominal muscles:** Inadequate abdominal muscle strength allows the pelvis to tilt forward, creating an exaggerated lumbar curve (lordosis). This large, strong muscle group responds to strengthening exercises like situps and crunches.
2. **Back muscles:** These muscles need to be strong and flexible, so both strengthening and stretching exercises are important. Upper-back muscles involve the shoulder and neck as well. Both upper- and lower-back muscles tend to become weak and tight unless regularly strengthened and stretched.
3. **Leg muscles:** Both the hip flexors (which pull the leg up toward your chest) and the hamstrings (back of the thigh) are important for a balanced posture. Inflexibility in these groups can pull the pelvis out of alignment. Strength in these groups is important for good antigravity support when standing, when getting up from a sitting position, and when lifting.

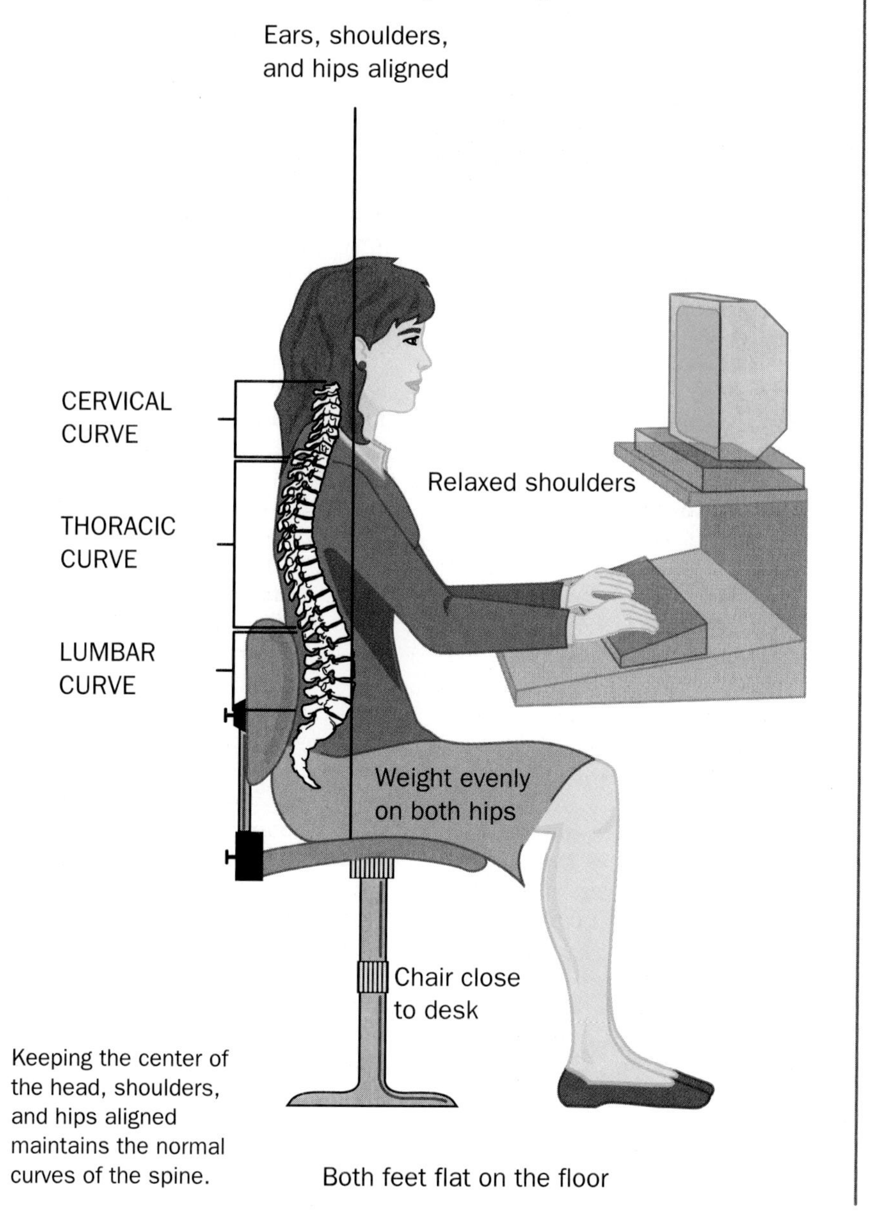

Keeping the center of the head, shoulders, and hips aligned maintains the normal curves of the spine.

ESSAY

STEPS TO HEALTHY FEET

8

The foot is an anatomical wonder. It has 26 bones, about 20 muscles, 33 joints, and more than 100 ligaments to hold it all together. This structure gives the foot an ideal combination of strength and flexibility to support the functions they must perform.

We take the structure and function of our feet for granted until they start to hurt. And even then we often continue to mistreat them, cramming them into shoes that are too tight, and then asking them to walk on concrete sidewalks and take us on long shopping expeditions. Extra body weight also means extra stress on our feet.

Most foot problems are preventable. The first step to happy feet is an understanding of their structure and function and how to use good footwear to support them in their work.

These feet were made for walking

Each time you take a step, your heel strikes the ground first. Then you roll over the ball of your foot onto your toes. Your arch flattens slightly as it absorbs the weight of your body. This foot continues to bear your weight until the heel of the other foot touches the ground. As you walk, your big toe maintains your balance while the other toes give your foot some resiliency. The outer metatarsals move to accommodate for uneven surfaces while the inner three stay rigid for support.

The most common cause of foot problems is ill-fitting shoes, which interfere with the foot's natural structure and function. The high heel is a case in point (80% of those suffering from foot problems are women). While many people think high heels look good and can be fun to wear, they should not be used for walking. Instead of the arches of the foot absorbing the force of the body's weight as you walk, all the weight falls onto the forefoot. This unnatural stress can injure the soft-tissue structures, joints, and bones.

Good shoes

Sensible shoes can prevent many foot problems and are especially important if you are doing any amount of walking. Fortunately, it is not uncommon nowadays to see both men and women wearing good walking shoes as part of their working attire or at least during their commute to work.

A good shoe has a sole that is strong and flexible and provides a good gripping surface. Cushioned insoles help protect feet from hard surfaces. Arch supports help distribute weight over a broader area, as do the arches in your foot.

Many people spend a great deal of time researching which brand of shoes to buy but do not spend adequate time evaluating whether or not the shoes are suited to their feet. A high-quality shoe is only worth buying if it fits! It's better to buy a lesser-quality shoe that fits well than a poorly fitting "high-quality" shoe. As one avid shopper put it, "If the shoe fits, buy it."

Shop for shoes in the late afternoon when your feet are at their largest. One foot is often bigger than the other; always buy for the bigger foot. The shoes you try on should feel comfortable immediately. Don't plan on shoes stretching with wear. The heel should fit snugly, and the instep should not gape open. The toe box should be wide enough to wiggle all your toes, and the

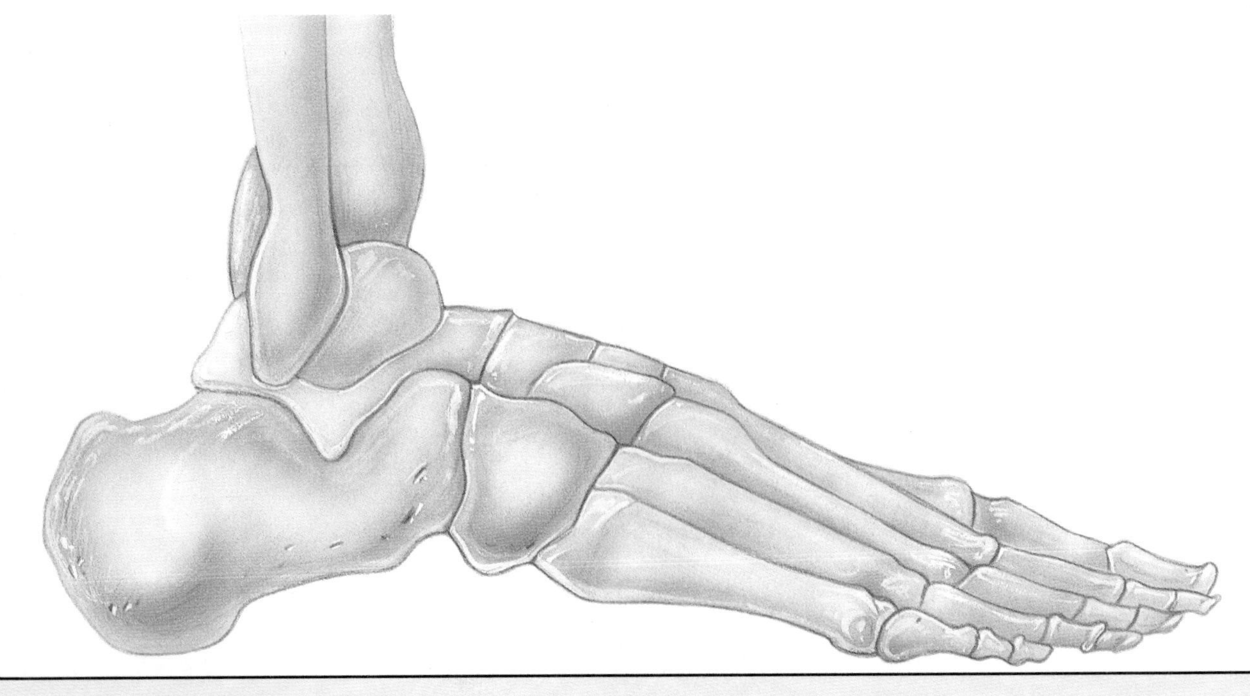

shoe should be as wide as your forefoot.

This last point is often overlooked. Many people think it is normal to push their feet into too-narrow shoes, perhaps believing this is part of the "support" shoes should offer. But unless you have problem feet, your arches give you good support, and shoes merely support this function, not so much the foot itself. If you do have problems with a foot structure that does not allow normal movement, a podiatrist can sometimes correct the problem with orthotics (shoe inserts) to change the way your foot works.

Buying shoes for children can be difficult because they may have difficulty deciding whether a shoe is comfortable and they may be more likely to go for fashion. The function of children's shoes is to protect developing feet. The heel should be snug, the sole should be flexible, and the toes should have room to move. The toe box should be about 1/4 in. longer, wider, and higher than the child's foot. When you place your thumb near the child's big toe, you should feel some room in the shoe.

Special editions

People who place extra demands on their feet should buy footwear accordingly. If you participate in some kind of sport, choose well-fitting shoes that will accommodate the extra demands placed on your feet. When you run or jog, your feet receive a force that is two to three times your body weight with each step. Running shoes are designed to give your feet extra cushioning for absorbing this shock, especially in the heel. A stiff heel counter and flared heel sole give good support and stability, and with the arch support help to prevent the foot rolling in or out. Padding on the inside of the heel helps to protect the Achilles tendon.

Walking shoes also provide support and shock absorption but are built to accommodate the rolling motion of your foot, rather than a pounding impact. Heel cushioning, arch support, and flexible soles are important. Walking shoes should also help prevent the foot rolling in or out.

Some activities, such as aerobic dance, require a repeated up-and-down motion. Aerobics shoes should have good overall support and adequate cushioning. Aerobic dance also requires frequent changes in direction, as do racquet sports and basketball. Shoes for all these activities should have good overall support to resist twisting of the foot and ankle. Running shoes are especially dangerous for activities requiring lateral movement, as you may trip over the flared heel.

Since good shoes are often expensive, we tend to hang on to them longer than we should. Usually the first thing to go in a sports shoe is its shock-absorbing ability, so the shoe may still look fine but not be performing one of its most important functions. Running shoes usually need to be replaced after about 400 to 500 miles, and aerobics and court shoes after about 50 to 75 hours of use.

ESSAY

LIVING WITH ARTHRITIS

9

A wellness life-style is not just for people who are already healthy. It's also for people with chronic diseases and disabilities. Limitation is a fact of life and something each of us learns to live with at some level. Limitation makes it even more important to live in a way that maximizes our potential for health and quality of life. The limitations imposed by arthritis provide an excellent illustration of this point.

There are over 100 different kinds of arthritis, and the disease affects one in seven people in the U.S. Except for infectious arthritis, which is caused by a specific disease agent, the causes of arthritis are not known. The symptoms vary in severity from day to day and even within a given day. People with arthritis often experience periods of remission. Since no cures are available, arthritis treatment means figuring out the best way to live with it. Education and self-care form the basis of arthritis treatment.

Living with arthritis

Self-care doesn't mean a person must "go it alone." On the contrary, self-care means knowing when to get professional help and working with health-care professionals to understand the nature of arthritis and treatment methods. The earlier the arthritis diagnosis, the better, since although there is no cure, medical and life-style treatment can still significantly reduce or delay its progression. The goal of arthritis medical care is to relieve pain, reduce inflammation, and prevent deformity.

Self-care means following the recommendations of health-care professionals and adapting one's life-style to one's physical limitations. The life-style modifications required by people with arthritis vary with the severity of their condition. Many people with arthritis continue to lead very active lives. Billie Jean King, the tennis player, is a good example. She has maintained a successful career as a professional player despite five bouts of knee surgery for osteoarthritis.

For people with more severe arthritis, coping with such simple tasks as preparing a meal or cleaning the house can provide real challenges. Some problems have simple solutions, like moving dishes to lower cupboards, using an electric can opener, and eating out more often. Relocating items and rearranging rooms can improve the comfort and safety of the home. People confined to a wheelchair will need to make more elaborate changes in their home environment. Self-care means managing one's resources in order to maximize quality of life.

Pain-relief techniques

Physical therapists play an important role in arthritis treatment. They educate the arthritis patient about various pain-relief techniques and therapeutic exercise programs.

Pain-relief techniques include hot and/or cold treatments, joint protection, and rest. Hot treatments include hot baths and showers, hot packs, heat lamps, electric heating pads and mitts, and paraffin wax. Cold treatments include ice packs and compresses. Both can decrease pain and improve joint mobility. Joint protection includes education about joint-sparing body mechanics. For example, people with arthritis in the wrist learn to push open doors with the side of their body, rather than the hand. Joint protection also includes appropriate use of orthopedic devices such as splints, walkers, and canes to reduce joint stress.

Rest is an important component of arthritis self-care. Prescribed rest may include complete bedrest, or periodic resting of affected joints, and/or stress-management relaxation techniques. Emotional rest includes participation in social groups and recreation. Arthritis can be a challenge to quality of life, and living with it requires a good attitude that can transform a potential invalid into an active family and community member.

Nutrition and weight control

Being overweight is a risk factor for the development of arthritis, and extra weight increases the stress on arthritic joints. Ten extra pounds on the torso translate into 40 extra pounds of force on the knee when standing. Weight control achieved through a well-balanced diet and a moderate exercise program can significantly slow the development of arthritis. Good nutrition is also important for maintaining general good health and well-being, which is as vital for the person with arthritis as it is for anyone else.

Exercise

For many years, medical scientists believed that osteoarthritis and other forms of arthritis were the result of "wear and tear" on joints and that exercise accelerated joint degeneration. Studies have shown that exercise does not appear to cause arthritis in healthy joints. Repetitive, high-impact movements such as running can speed the progression of arthritis in already damaged joints, however. Unfortunately, because of the confusion that has surrounded the arthritis–exercise issue, many people with arthritis have avoided all but the very mildest forms of exercise. Many people with less severe arthritis have unnecessarily restricted aerobic activity for fear it would worsen the disease. But a lack of physical activity actually accelerates joint degeneration and worsens arthritis symptoms. A sedentary life-style leads to loss of muscle strength and low fitness levels, which make movement even more painful and difficult, leading to further restrictions in activity, and an even greater decline in fitness. When muscles and joints atrophy, the resulting weakness makes joints even more unstable. Recent studies have demonstrated that aerobic activities, especially those that support body weight such as cycling and swimming, can be appropriate for people with osteoarthritis and rheumatoid arthritis who have fairly good joint mobility; can increase aerobic capacity, muscle strength, and functional status; and can improve pain tolerance, mood, and quality of life.

Physical therapists help educate arthritis patients about therapeutic exercises for the maintenance of joint function. These typically include range-of-motion (ROM) exercises to increase joint mobility and resistance exercises to increase the strength of muscles, tendons, ligaments, and other tissues that are part of the joint structure. Many physical therapists recommend exercises performed with elastic tubes to provide resistance (see illustration). Flexibility and strength help protect joints from stress.

Arthritis self-care means learning to balance exercise and rest. Too much exercise can lead to pain and inflammation, while too much rest can cause joint stiffness. Exercise must often be done in very small amounts several times during the day if arthritis is severe. If pain persists for an hour or more after exercise, the patient has overdone it and must reduce activity to a lower level.

Physical therapists often prescribe exercises with elastic bands for increasing strength and flexibility. The bands provide a firm but gentle resistance without causing joint stress. In this exercise, which strengthens the muscles of the arms and shoulders, the patient holds the arms out straight, and then pushes down with one arm while pulling up with the other. The position is held for 10 seconds. The exercise is then repeated in the other direction.

ESSAY

10 FLEXIBILITY: TAKING IT TO THE LIMIT

Unless you were in physical education class a long time ago, your instructors probably warned "Don't bounce!" whenever you assumed a stretching position. But ask your parents about physical education classes and sports programs and how they warmed up for activity. They might remember doing ballistic stretches, where muscles were forcefully pulled then released. If you pull harder on a muscle, it will stretch farther, right?

Not so. As scientists have come to understand the physiology of muscle tissue, more effective methods for increasing flexibility have been devised.

Flexibility: a question of degree

Flexibility refers to a joint's range of motion (ROM). Range of motion is the maximum ability to move the bones of a joint through an arc. For example, if an injured knee can go from a 50° angle fully flexed to a 120° angle fully extended, the knee's ROM is 70°. Physical therapists measure improved joint mobility by an increase in its ROM.

Many athletes know the importance of flexibility. Some sports such as dance, gymnastics, figure skating, and diving require a great degree of flexibility. But even athletes in other sports not noted for their flexibility requirements often include some sort of stretching exercises in their workouts to increase or maintain flexibility. Flexibility improves performance and prevents injury. Runners, for example, use stretching exercises to prevent muscle tightness in the lower back and hamstring muscles and to improve stride length and overall running form.

Some degree of flexibility is necessary for everyone, simply for the performance of daily tasks. Limitation in your shoulder's ROM can make it difficult to pull a shirt on over your head, lift a dish from a shelf, or even brush your hair. As we age, muscles tend to shorten, especially if we do not do stretching exercises regularly. Shorter muscles decrease a joint's ROM. This may be part of the reason age is associated with stiff joints. It's important to remember, however, that this stiffness is due more to lack of exercise than to merely getting older.

Mind and muscle

The first step to take in increasing flexibility is to relax. Sounds simple, right? But if you ever visit an exercise class, you'll notice some people who are all tense, rigid, and hunched up, because the stretching position is uncomfortable and the muscles tighten up in protest. These people figure they'd better push a little harder, and they tense up even more. This rigid stretch is not accomplishing much.

When the muscle is stretched too fast or too hard, a sensory organ located in the muscle, called a muscle spindle, senses the stretch and sends a message to the muscle tissue to contract. This particular action is called the *stretch reflex.* So to stretch effectively, you've got to stretch in a way that doesn't alarm the muscle spindles. One of these ways is called *static stretching.*

A good static stretch is slow and gentle. You get into a stretching position that is not uncomfortable and relax. Then, continuing to relax and breathe deeply, you reach just a little farther . . . and a little farther, holding the stretch for at least 30 seconds. If you have difficulty relaxing, you know you have stretched too far. Ease up until you feel a stretch but no strain.

Often after holding a stretch for 10 seconds or more, you can feel the muscle relax even more, which results in a greater increase in flexibility. This response is called the *inverse stretch reflex.* Sensory organs in the tendon respond to increases in tension, produced by either contraction or stretching, by telling the muscle to relax in order to prevent injury.

Physical therapists have designed some special stretching procedures that go beyond static stretching and take advantage of our understanding of muscle physiology. These techniques are called *proprioceptive neuromuscular facilitation,* or PNF for short.

One PNF technique is the contract–relax method; the goal is to activate the inverse stretch reflex by first contracting the muscle to be stretched. The stretch is slow and gentle, so the muscle spindles keep quiet. Try this: First stretch your neck by moving your head to your left shoulder. Keep your shoulders relaxed. Note how far you can go. Hold that position but put your left hand on the right side of your head and push against your hand to make an isometric contraction (muscle tension but no movement). Hold for 5 to 10 seconds. Now with the same hand gently pull your head a little closer to your shoulder. Be sure not to pull too hard. Did your stretch increase? This is an exercise physical therapists often teach to patients with neck and upper back tension.

The contract–relax with agonist-contraction (CRAC) method starts out the same as the contract–relax method, but then you also apply resistance to the muscle group opposite to the one being stretched. In general, when one muscle group contracts, the group opposite to it relaxes; this is called *reciprocal inhibition.* Try it: After you've done the contract–relax sequence above, move your left hand to the left side of your head and push, without moving your head, creating tension on the left side of the neck. Hold for 5 to 10 seconds. Now relax and pull very gently on the right side again, moving the head a little farther toward the left shoulder. (See illustration.) Compare the flexibility on the right side of your neck with that on the left. Any difference?

Versions of PNF stretches using partners are sometimes taught, but it is extremely important that the partner not force the stretch too far. The best partner is yourself, a trained instructor, or a physical therapist.

Stretching it

When a relaxed muscle is physically stretched, its ability to elongate is limited by connective tissue structures, such as fasciae. Collagenous tissue stretches best when slow, gentle force is applied at elevated tissue temperatures. An external source of heat such as hot packs or ultrasound can be used. But aerobic exercise is a more effective way to raise muscle temperature, if possible. That's where the term *warm-up* comes from. It's important to warm up *before* stretching, not vice versa. Stretching cold muscles is less effective and may even cause injury.

Example of PNF technique for neck muscles.
1. Move your head to your left shoulder.
2. Put your left hand on the right side of your head, and hold an isometric contraction for 5–10 seconds.
3. Move hand to other side of head, and hold an isometric contraction for 5–10 seconds.
4. Relax, and gently pull head toward shoulder.

ESSAY

11 STRENGTH TRAINING: ANTIDOTE FOR AGING

Some people still associate strength training with Olympic musclemen, Charles Atlas, and other extremely hypertrophied athletes. Few see themselves in these models, and the idea that strength training might have benefits beyond the development of a superhero body never occurs to many people. This is unfortunate, because almost everyone can benefit from a well-structured strength training program.

Why weight?

Weight training is a form of resistance training, which simply means any kind of exercise in which the muscles exert force against a resistance. Weight training uses weight machines and free weights to apply resistance. Resistance can also be applied with rubber tubes, another person, or even water. One of the biggest advantages of weight training is that resistance can be applied in a measured, progressive fashion. After your strength increases enough to lift three blocks of weight easily, you add another block.

Certainly the most well-known benefit of weight training is an increase in muscle size and strength. But the benefits of weight training go beyond the acquisition of muscular prowess and read like an antiaging potion. In our sedentary society, many orthopedic problems are the result of weakness and inflexibility, which are often shrugged off and attributed to the aging process. Much of this loss of physical function, however, is due to inactivity and a consequent decline in physical fitness, rather than to aging.

Another important result of weight training is that other connective tissue structures, such as tendons, ligaments, and joint capsules, increase in strength as well. Weight training seems to increase bone strength, too, and may thus help maximize deposition of bone mineral in young adults and prevent, or at least slow, its loss in later life.

People trying to lose weight are usually told to participate in aerobic exercise, because it burns more calories than weight training. While aerobic exercise requires a greater expenditure of energy per minute, weight training can still be a helpful supplement to aerobic exercise for two reasons. First, weight training may help prevent injuries, especially for people who have a low level of fitness. Many injuries that occur in new exercisers are due to pushing a deconditioned body to do too much, too soon. The cardiovascular and respiratory systems of a new exerciser are often ready to do more than their muscular and skeletal systems. A new exerciser can benefit from "getting into shape" before beginning an exercise program.

The second reason weight training can play an important role in a weight control program is that weight training will help preserve or even increase muscle tissue mass. Since metabolic processes occur at a much faster rate, even at rest, in nonfat body tissues, the greater your fat-free mass (a large proportion of which is muscle tissue), the higher your metabolic rate. The higher your metabolic rate, the more calories you can eat without gaining weight. This preservation of fat-free mass is another antiaging property of weight training, since a decline in muscle mass is one of the main reasons metabolic rate decreases as we age.

Perhaps one of the greatest benefits of weight training is that it may help prolong independence in older adults. As we age we often lose our ability to live independently because we have lost the basic strength to climb stairs, carry groceries, get up after a fall, or even stand up from the toilet. Sometimes injury or illness leads to a period of bedrest and/or reduced activity, which precipitates a sharp decline in physical fitness. Recent studies have provided a dramatic illustration of the important contributions weight training can make to the quality of life in very old people. Men and women in their eighties and nineties who participated in a weight training program made substantial gains in muscle strength, with strength increases of 30% in some cases. This improvement in strength significantly increased the ability of these elders to perform daily tasks. In one case, a woman was able to walk without a cane for the first time in many years.

Other health and fitness benefits

Some research studies have documented a small increase in cardiovascular fitness (aerobic endurance) in subjects who lifted weights, especially those who participated in a circuit training program. Circuit weight training involves high-volume, low-resistance lifting, moving quickly from station to station. And as in any other type of activity, the greatest fitness benefits accrued to those with the lowest fitness levels.

Regular weight training may also help to improve blood lipid profiles, increasing the "good" kind of cholesterol (HDL cholesterol) and decreasing total and the "bad" kind of cholesterol (LDL cholesterol). Some studies have also documented improvement in other cardiovascular risk factors, such as blood sugar regulation and blood pressure.

Importance of an individualized program

Since weight training has traditionally been perceived as a very demanding form of exercise, most people have assumed that it is appropriate only for the already athletic. But fortunately, recent studies, such as the research with old adults described above, have shown that even people with limited physical ability can benefit from an appropriate weight training program. The key is matching the weight training program to an individual's current fitness level and health concerns.

People with hypertension have often been told to avoid lifting heavy weights because of the transient rise in blood pressure that occurs during lifting. But research with borderline-hypertensive subjects who practiced correct breathing procedures (no breath holding) found that circuit weight training with low-resistance lifting did not elevate heart rates or blood pressures to dangerous levels.

If you've ever been to a physical therapist for an injury, chances are part of your treatment involved exercise: strengthening and stretching the affected areas to restore function. Working with weights is often a part of the exercise prescription. Many people with orthopedic problems, including those with arthritis, have found that the gains in strength and flexibility from regular participation in a weight training program have greatly improved their functional capacity.

While it is important for anyone new to weight training to receive instruction by a qualified professional, it is essential that anyone with any kind of limitation seek the guidance of a health professional as he or she designs an exercise program.

ESSAY 12

THE CONSTRICTION OF ADDICTION

> I had a drug problem that was really a "me" problem, and I wanted to blame cocaine. People always talk about drug abuse, but I don't know of one drug that has ever been abused. It's self-abuse. It's not the vehicle, not the coke. It's you. You abuse yourself.
>
> *Eugene (Mercury) Morris, former pro football player and former cocaine addict*

Party time. Laughter and music fill the house. In one room, a small group of high school students huddles in a circle. The air tingles with excitement and expectation. One student carefully fills the bowl of a small pipe with some porcelain white crystals. A match is lit. Crackle . . . pop. The pipe is passed around.

For most of the students in this group, their experimentation with crack cocaine will continue harmlessly for only a few episodes. But the lives of one or two in this group will become so hopelessly enmeshed in the self-destructive behavior surrounding drug abuse that their personalities will bear the scars for years to come.

What is addiction?

To become addicted is to devote or surrender oneself to something habitually or obsessively. It is somewhat ironic that the word *addiction* is derived from the Latin verb meaning "to give assent," since we tend to think of addicts as being incapable of giving assent, having lost control of the habit in question.

We commonly use *addiction* with reference to drugs, including familiar substances such as nicotine, alcohol, and caffeine, as well as pharmaceutical preparations both legal (sleeping pills, diet pills, antihistamines) and illegal (cocaine, heroin, marijuana). But a person can also become addicted to behaviors, such as exercise, the adrenalin rush of taking risks, abnormal eating rituals, destructive relationships, work, and probably even to being a student.

Addiction means more than having a strong desire for something. The experience of addiction is a negative one, which distorts reality and limits participation in life. Addiction is marked by a gradual reduction in awareness, a decline in self-esteem, and decreased involvement with other people and activities. Although people initially become addicted to something because it feels good, addiction is not pleasurable once the initial "high" has passed, and addicted people usually do not feel good most of the time, which results in increased cravings for something to fix the problem.

Physiology of addiction

Addiction is not simply a state of mind. It often has a strong physiological component. Animal studies have shown that many drugs, including cocaine and nicotine, have strong reinforcing properties. If permitted, some laboratory monkeys will continually reinject themselves with cocaine, while forgoing food, water, sleep, and sex. Why do some drugs feel so good?

Many drugs exert their effects by modifying the transmission of nerve impulses. These modifications lead to altered physiological and psychological states. Cocaine is a good example. It blocks the reuptake of the chemical messengers (neurotransmitters) norepinephrine and dopamine, so that they accumulate at the receptor sites, causing increased stimulation of the nerve cells. The short-term effects are increased stimulation of the sympathetic nervous system, which results in increased heart rate, blood pressure, blood sugar, and body temperature. Modification in the transmission of nerve impulses may also be involved in producing the euphoria associated with cocaine use.

Addiction to cocaine may also result from the long-term effects on nerve transmission. After a week or more of regular use, some nerve cells seem to become depleted of the neurotransmitter dopamine. Scientists have theorized that this depletion may be responsible for the strong physiological cravings that characterize cocaine addiction.

While the physiological effects of drugs are very important, they are only part of the explanation of why people become addicted to a given substance. Why do some people use drugs occasionally with no disruptive physical or psychological effects, while others just can't say no?

Some researchers theorize that a genetic component is involved, since substance abuse tends to run in families. The effects of a given drug vary from person to person, and genetic differences could explain this variation in individual response. Other researchers argue that differences in drug use are due to family environment rather than genetics. The believe that substance abuse leads to dysfunctional families and inappropriate role models for the children, who in turn become prone to addiction.

Research on drug addiction has also suggested that part of addiction is situational. For example, people may use heavy doses of morphine during a painful illness, but then be able to give up the drug once they have recovered and the drug is no longer needed. Similarly, some young men who served in the military during the Vietnam war were heavy heroin users while overseas, but gave up the drug once they were back with their families.

Certain personality traits may also predispose people to addiction. People prone to addiction are more likely to have low self-esteem and less self-reliance. They are more likely to look to other people, behaviors, and substances, rather than to themselves, for solutions to problems. Stressful situations often exacerbate addictive behavior. Boredom, loneliness, shyness, desire for peer approval, and a perception that life lacks meaning can all contribute to a desire for something that will take you out of yourself and your rut.

Predicting addiction?

Although many treatments exist, problems with addiction are often very difficult to cure. People who are able to recognize the beginnings of addictive behavior in their own lives and seek help to change this behavior in its early stages generally have an easier time overcoming addiction.

How do you know if you're addicted? Even a single "yes" to any of the following questions indicates that the behavior may become a problem and possibly an addiction.

Does the behavior provide the primary source of gratification in my life?

Does it provide the primary means of escape or avoidance of problems?

Does it decrease my self esteem?

Am I developing tolerance to the substance, needing more than I used to to achieve the desired effect?

Have I been unsuccessful in my attempts to stop this behavior?

Do I need it to function?

Is it causing (will it cause) the gradual development of health problems?

Has this behavior resulted in suggestions from others to change or stop?

Does this behavior occur as a predictable, ritualistic, and compulsive activity?

Wellness: addiction prevention

Developing a wellness life-style, based on self-responsibility and self-love, is one of the best ways to prevent addiction. A wellness life-style means cultivating healthy pleasures that replace the need for unhealthy addictions. A life with purpose, meaning, and values that offers fulfillment and enjoyment is more resistant to addiction.

ESSAY

13 SPECIAL NEEDS: ACCESS TO WELLNESS

People with special needs: It's a group that many of us are a part of, and that many more will join at some time in the future. It has already been emphasized that a wellness life-style doesn't refer to a certain set of achievements, such as being able to run a given number of miles, consume a diet with only so many grams of fat, and never feel stress about upcoming exams. Rather, a wellness life-style means living in a way that maximizes your own potential for health and well-being. People with special needs can benefit from a wellness outlook and life-style as much as anyone else.

What is meant by special needs?

A person with a disability needs something extra to function the way able-bodied people do. For example, a person with a spinal cord injury of nerves to the legs has special mobility needs, usually needing a wheelchair to get around. The phrase *special needs* is meant to be more inclusive and a less value-laden label than *handicapped* or *disabled*.

A few examples of conditions requiring special accommodations include the following: mental retardation; serious emotional disorders, such as schizophrenia and depression; learning disabilities; hearing impairment; visual impairment; neurological and muscular handicaps including cerebral palsy, spinal cord injury, and spina bifida; and orthopedic handicaps such as amputations and congenital limb deficiencies. People with chronic illnesses including AIDS, respiratory problems, cardiac conditions, kidney disease, diabetes, and arthritis may also have special needs, depending on the type and severity of their condition. Temporary conditions such as pregnancy and disabilities such as broken bones and torn ligaments may also require special accommodations.

The "special needs" of these conditions are as varied as the conditions themselves. Consider the college environment. Students with mild learning disabilities may need modification of the traditional testing structure. Some may require more time to complete examinations because of a slower rate of reading and writing. The quality of the finished work may be as good as that of other students, but it may simply take longer to complete. While they might fail an exam in a traditional timed setting, they perform well when given adequate time. On the other hand, someone with visual impairment may require access to audiotapes of their texts and the assistance of a reader. Meeting such special needs allows people with disabilities to adapt to situations and environments that would otherwise be less accessible.

Maximizing health: access to medical care

Improvements in physical health almost always mean improvement in quality of life. Modern medicine continues to expand its ability to treat many limiting conditions. Everyone needs good medical advice and health education about how to maximize one's health through medical therapy and appropriate self-care measures.

Many people with disabilities need long-term access to special care and/or certain types of therapy. Physical therapists, occupational therapists, psychotherapists, and other allied health professionals make significant contributions to the quality of life for people with special needs. Quality of life may also be improved with well-designed therapeutic equipment, such as wheelchairs, prostheses, and hearing aids. Guide dogs are a tremendous help for the visually impaired.

Physical access to daily life and recreational activity

Much can be done to improve the ability of people with special needs to live independently. Conceptually simple measures (not so simple economically) such as making a town more "wheelchair friendly," manufacturing automobiles adapted to a variety of physical limitations, and adapting sports and recreational equipment can help people with special needs become more independent and increase the activities in which they can enjoyably participate. Rewarding occupations, participation in one's community, and satisfying recreational pursuits all require some form of physical access.

People with severe limitations are the most in need of technology that can improve their ability to perform routine daily tasks, such as eating, dressing, and even speaking. Stephen W. Hawking, world-renowned physicist, writer, and a person living with amyotrophic lateral sclerosis, a progressive disease of the motor neurons, uses a special communications program on a small computer equipped with a voice synthesizer, all mounted on his wheelchair, to write and speak. Without this special equipment Hawking's eloquent ideas would be unable to find expression.

Intellectual wellness: access to education

National legislation continues to promote improved educational opportunities, especially for children and adolescents with special needs. One of the often misunderstood concepts mandated by such legislation is mainstreaming. *Mainstreaming* is not simply integrating children with special needs into the traditional classroom, although this is how the term is often used. The goal of mainstreaming is to place each special needs child in the "least restrictive environment" in which educational and other needs can be satisfactorily met. For some children, mainstreaming may mean moving into a regular classroom, perhaps with some special support. Placing individuals in programs for which they are not ready is recognized as being inappropriate for the program, and not in the best interests of the special needs individual. A certain percentage of special needs students will always require special programs of some sort.

Personal best: the wellness challenge

For everyone, including special needs people, the key to high-level wellness is to focus on the potential, not the limitation. A wellness life-style means finding meaning in life and developing an ability to get the most out of each day.

Meaning can be found in many ways. Helping others is a particularly rewarding experience. Pursuit of some sort of craft or art form gives us a sense of accomplishment. Meeting the intellectual challenge of an educational degree is a milestone for many.

Many people with special needs find emotional release, fun, and satisfaction from participation in physical activity and sports. Much sports equipment has been adapted for specific limitations. Bicycles, kayaks, and skis have been adapted for paraplegics and amputees. Wheelchair basketball, baseball, tennis, and even dancing are becoming more popular. There are wheelchair divisions for many marathons. While some people achieve international recognition, meaningful involvement also occurs at even the most amateur level.

ESSAY

EXERCISE HIGH

14

Self-reports from fitness enthusiasts of improved mood during and following exercise were once dismissed as the crazy delusions of "health nuts." Before the North American fitness boom began in the 1970s, few people recognized the health benefits of regular physical activity, and anyone who ran without being chased was considered suspect.

As more people began to participate in some type of exercise program, these self-reports of mood enhancement, often bordering in character on testimonials, continued to be heard. Fitness instructors noted that many people initially began to exercise because they thought it would help them lose weight, or be good for their health, but would continue because, they said, exercise made them feel so much better.

As psychologists began to investigate this topic, they found that when questioned, most people report that they feel good and have less tension and anxiety after vigorous exercise. Researchers have found that tension and anxiety are generally both reduced for about 2 to 5 hours following vigorous exercise. Continued participation (6 to 20 weeks) in regular physical activity seems to produce long-term changes as well, at least in some groups of people. In particular, people with depression or low self-esteem often experience improvement following exercise program participation.

Runner's high

Some people report feeling "better than good" when they exercise. These accounts of euphoria and states of consciousness similar to those described by people using drugs such as heroin led to use of the term *runner's high,* since these descriptions first came primarily from long-distance runners. An example of such a description comes from A. J. Mandell, in a paper titled "The *Second* Second Wind":

> Thirty minutes out, and something lifts. Legs and arms become light and rhythmic. . . . The fatigue goes away and feelings of power begin. I think I'll run twenty-five miles today. . . .
>
> Then, sometime into the second hour comes the spooky time. Colors are bright and beautiful, water sparkles, clouds breathe, and my body, swimming, detaches from the earth. A loving contentment invades the basement of my mind, and thoughts bubble up without trails. . . . The running

literature says that if you run six miles a day for two months, you are addicted forever. I understand. A cosmic view and peace are located between six and ten miles of running. I've found it so everywhere.

After the run I can't use my mind. It's empty. Then a filling begins. By afternoon I'm back into life with long and smooth energy, a quiet feeling of strength, the kind wisdom afforded those without fear, those detached yet full.*

Such descriptions have intrigued both exercise scientists and the lay public and have raised several questions that so far remain only partially answered. The first concerns the threshold for this exercise high. Does the exercise have to be of high intensity and/or long duration? How high and how long? Reports such as Mandell's suggest that the activity does have to be at least moderately vigorous and at least 30 minutes long before even a mild high is perceived. Studies show that low-intensity, short-duration forms of physical activity can also improve mood and allay feelings of anxiety and depression, but are generally not associated with feelings of euphoria.

Why high?

Other questions concern the psychophysiological explanation for these feelings. Is the runner's high similar to highs perceived during drug use? And does it result from similar changes in brain biochemistry? Are these or similar changes also responsible for those reports of "feeling good although not exactly high" following exercise?

*Reproduced with permission of Slack, Inc. (Mandell, A. J., The *second* second wind. *Psychiatr. Ann* 9:57–68, 1979).

As scientists have come to understand something of brain biochemistry, some interesting hypotheses have emerged. The most publicized of these has focused on a group of chemical messengers found in the central nervous system (brain and spinal cord) called *opioids,* since they are similar in structure and function to the drugs that come from the poppy flower: opium, morphine, and heroin. Endorphins and enkephalins belong to this group. They not only inhibit pain, but seem to have other roles in the brain as well, such as aiding in memory and learning and registering emotions. It has been suggested that the release of endorphins and other endogenous (produced by the body) opioids is responsible for the exercise high. The popular imagination has been captivated by the idea that something as simple as exercise could cause the production of these "endogenous drugs." Some felt this hypothesis would also that facilitate the release of sugar from glycogen stores in the muscles and liver, bringing the blood sugar up into normal range. Homeostasis is regulated by the nervous and endocrine systems, which allow all of the body's systems to communicate with oneur. The most easily measured variable is plasma endorphin level, but this variable does not appear to reflect endorphin levels in the brain. Research using plasma endorphin levels has generally failed to show a relationship of this variable to either exercise intensity or psychological state.

Animal research has suggested endogenous opioids increase with level of exercise. Recent research has suggested that endogenous opioid concentrations increase when certain nerve fibers are activated, either through repetitive muscle contraction (such as that experienced with exercise) or acupuncture.

The endogenous opioids may help the body recover from prolonged exercise, as they seem to enhance mechanisms important during this period: raising pain threshold, slowing heart rate, decreasing blood pressure, and enhancing parasympathetic outflow (which leads to relaxation) while inhibiting sympathetic activity (the "fight or flight" response).

Other biochemical explanations

Some research has suggested that changes in the concentration of monoamines, another group of chemical messengers found in the central nervous system, may be responsible for the changes in emotional state observed with exercise. In particular, norepinephrine (NE) and serotonin (5-HT) concentrations have been shown to change with exercise, at least in animals. Since abnormal levels of these chemicals have been associated with depression in humans, it has been speculated that the antidepressant effect of exercise may involve improving regulation of these substances in the brain. It is also possible that the endogenous opioids interact with monoamine regulation in some synergistic fashion to produce euphoria and mood improvement.

Distraction hypothesis

Some psychologists have suggested that the positive effect of exercise on mood is similar to that produced by many other relaxation techniques including yoga, biofeedback, and meditation (see Essay 17). These researchers argue that exercise provides a distraction from sources of daily stress, and in this way helps the exerciser to feel better and more relaxed.

ESSAY

IN SEARCH OF MORPHEUS

15

While those who get it usually take it for granted, people who don't would do almost anything for it. Although it's extremely valuable, you can't buy it. You can't borrow it, steal it, or give it away. And the harder you try to get it, the less likely you are to succeed. What is it? A good night's sleep.

Almost everyone has trouble sleeping occasionally, and experiences nights during which they toss and turn, look at the clock, get upset, and toss and turn some more. At any one time, about 15 to 20% of adults in North America complain of problems sleeping. Fortunately, there are no serious consequences to missing a good night's sleep once in while, although even a few sleepless nights can dampen one's good humor and interfere with one's mental alertness (students who pull all-nighters take note!).

Sleep problems often last for a relatively short period of time and go away on their own even without treatment. Some people, however, suffer from chronic insomnia, or difficulty sleeping, that may persist for months or even years. Chronic sleep deprivation can seriously impair one's mental and physical health and ability to function effectively.

Insomnia

Insomnia may include any or all of the following problems:

1. Taking a long time to fall asleep.
2. Awakening frequently during the night.
3. Awakening too early in the morning.
4. Feeling tired and dissatisfied with one's sleep upon awakening.

Insomnia that continues for more than a few weeks is a problem requiring medical attention because occasionally it may signal a more serious health problem. One of these is sleep apnea, a potentially dangerous condition that occurs when the sleeper stops breathing for about 30 seconds and awakens, snoring loudly. Sleeplessness is sometimes a side effect of medication. Overuse of caffeine, alcohol, sleeping pills, and other drugs can also cause insomnia.

Why not take a sleeping pill?

Sleeping pills are among the most commonly prescribed drugs in North America. An occasional dose of sleep medication may be helpful when taken as directed, but, in general, medication only worsens the problem. Sleeping pills disturb the sleep cycle, so sleep is less satisfying even though you may get more of it. They often leave the user with a hangover that results in daytime fatigue. The user quickly builds up a tolerance to the medication so that it becomes less effective within a week. One of the biggest dangers of sleeping pills is addiction, which can be very difficult to overcome.

Some sleeping pills can even *cause* insomnia by suppressing the brain's production of dopamine, a neurotransmitter that helps you go to sleep. Sleeping pills pose a danger of overdose, especially if taken with alcohol or other drugs. Many people believe alcohol will help them relax and go to sleep, but like sleeping pills, alcohol disrupts the sleep cycle. While alcohol may help you fall asleep, it usually produces light, restless sleep, and the person often awakens suddenly during the night, unable to go back to sleep.

The biggest problem with sleeping pills is that they do not address the real causes of the sleeping problem: bad habits and stress.

Sleep therapy

While it is normal for some insomnia to occur during times of stress, it should disappear once the situation is resolved. When it continues it is often because the person has developed a poor sleep environment and/or poor sleep habits.

Sometimes the sleep environment is the problem. The sleep environment should be comfortable, restful, and associated with relaxation and sleep. Sleep and work areas should be separate. Changing simple physical characteristics of the sleeping area often improves sleep quality. Most people sleep best when the room temperature is between 60° and 65°F. Noise level can be a more difficult problem to control. White-noise machines or tapes can provide a soothing sound that helps cover traffic and other noise. Some people even resort to earplugs. Shades that block light can help darken rooms with windows near street lights or keep out the early morning sun.

Regular use of stimulants can interfere with one's ability to fall and stay asleep. A decrease in the consumption of caffeine (found in coffee, tea, and cola drinks) often improves sleep quality. The nicotine in cigarettes is a stimulant and should be avoided. Heavy smokers experience more sleep problems.

A large meal before bed can inhibit sleep. A light snack, however, can help one sleep better. Exercise helps decrease muscle tension and improve sleep quality. Exercise can also improve one's ability to manage stress, to feel less worried and more in control. But beware: exercise too close to

bedtime can wind you up instead of down. Sleep experts generally recommend exercising in the afternoon. If an exercise program is making sleep worse, you are probably overdoing it.

Sleep comes more easily to those who go to bed with a peaceful mental attitude. It is helpful to relax for at least an hour before bed. Read, listen to music, knit a sweater, take a warm shower. Avoid activities that wake you up. A pre-bed routine helps get you ready for sleep.

If sleep is still elusive, one can try a more rigorous approach called *stimulus control,* which consists of behaviors designed to help a person associate going to bed with sleep and sleepiness. Examples of instructions for stimulus control include the following. Lie down to sleep only when sleepy. Do not use the bed for anything but sleep (and sex). Don't read, watch TV, study, or eat in bed. If you haven't fallen asleep after 10 minutes, get up and do something else until you get sleepy again. Troubled sleepers are also advised to get up at the same time every day and not to nap during the day.

Stress management for insomniacs

If psychological stress is the cause of insomnia, stress-management training programs can be beneficial. Professional counseling with someone familiar with sleep disorders can be especially helpful. Resolving conflicts and dealing with problem situations are better than worrying. Relaxation techniques (see Essays 17 and 18) usually help the insomniac achieve the before-bed relaxation conducive to a good night's sleep.

ESSAY

16 THE EYES (AND EARS) HAVE IT

Think for a moment about the importance of sight and sound. Consider your continual reliance on your vision and hearing for information from your environment. Reflect on the pleasures of perception: the faces and voices of friends and family; a quiet country evening with crickets chirping, fireflies flashing, and a sky of Maxfield Parrish blue, accented with the last rosy rays of sunset.

Like good health, we take our special senses for granted. While some disorders involving the eyes and ears are not preventable, many factors affecting the health of these sensory systems are under our control. Both accidental injury and chronic abuse take their toll on the health of our eyes and ears.

Prevent hearing loss

Some things improve with age, but hearing is not one of them. Damage to the hair cells that convert sound waves into nerve impulses accumulates over a lifetime, and by the time hearing loss is discovered, irreversible damage has already occurred. The most common form of hair cell wear and tear is exposure to excessive noise. Damage increases with both the intensity and duration of exposure. The hair cells appear to be less traumatized by short periods of loud noise than by chronic exposure to moderately loud noise.

Excessive noise can be found both at home and in the workplace. Workplaces must abide by certain safety standards to limit workers' noise exposure, either by changing working conditions or supplying hearing protection. Common sources of damaging noise at home include engines (motorcycles, snowmobiles, motorboats, and lawnmowers); power tools (table saws, chain saws); firearms; and loud music, including car stereos, radio headphones, and rock concerts.

What is excessive noise? Noise need not reach the pain threshold to be damaging. Eight hours of noise at the level of most vacuum cleaners is believed to damage hearing. Hearing loss can be prevented, or at least postponed, by avoiding excessive noise and wearing hearing protection whenever necessary.

Prevent eye injury

Eye injury can occur in several ways. Common causes of eye injury include flying particles and dusts, chemical splashes, and radiant light. Studies show that 90% of eye injuries occurring in the workplace are preventable with protective eye wear. Protective eye wear should also be worn at home for tasks that could lead to eye injury, such as trimming hedges and using power tools.

Eye injury can also occur when playing sports. The most common sports eye injuries are usually caused by the impact of a ball. Ball speeds in squash, racquetball, and baseball often reach 75 to 125 mph, which can cause serious trauma. Eye injury can also result from getting struck with a racket, hockey or lacrosse stick, golf club, or such; and from collision and body contact. Wearing appropriate eye protection can greatly decrease the likelihood of eye injury.

Prevent sun damage

The ultraviolet (UV) rays of the sun appear to age the eyes, just as they age the skin (see Essay 4). Two eye diseases associated with aging are cataracts, in which the lens of the eye loses its transparency, and senile macular degeneration (SMD), in which new blood vessels grow over an area of the retina called the *macula lutea,* reducing visual clarity. People with light-colored eyes are most vulnerable to UV damage. Glasses that filter out UV light protect eyes from these damaging rays.

Smoke gets in your eyes

People who smoke have higher rates of cataracts. Researchers estimated that about 20% of all cataracts may be attributed to cigarette smoking. The way in which smoking increases cataract development is unknown, but scientist speculate that it may increase free-radical damage to the lens.

Prevent eyestrain

While it has not been conclusively demonstrated that prolonged periods of close work, such as reading and working at a computer (the academic life-style!), can cause nearsightedness, it can cause headaches and eyestrain. When involved in close work, look up frequently and gaze into the distance to prevent eyestrain and relax the ciliary muscles, which adjust the lens of the eye for near or far vision. Reading and watching television in a dark room and working with glare on reading material and computer screens should be avoided.

Screening tests

Early diagnosis of eye and ear disorders allow one to benefit from early treatment that may prevent or at least retard further damage to vision and hearing. Routine eye exams use simple and relatively painless procedures to detect cataracts, glaucoma, SMD, changes in the blood vessels of the eye, and other disorders. Hearing tests can pinpoint specific areas of hearing loss, which may be partially compensated for with some sort of hearing aid. Early detection of hearing loss can also help one take action for the prevention of future hearing loss.

Good nutrition and the antiaging life-style

When asked what it is like to grow old, someone once replied, "Smear some Vaseline on your glasses and stuff some cotton in your ears, and you'll see what it's like." While no one has discovered the fountain of youth, there are a few things we can do to slow the physiological changes that occur as we age.

Loss of hearing and visual acuity are sometimes caused by artery disease (atherosclerosis), which is a long-term degenerative process affected by a number of variables, some of which are under our control. Regular exercise, avoiding cigarette smoke, and following a "heart-healthy" diet (see Essay 20) can help prevent atherosclerosis as well as other risk factors that contribute to it, including abdominal obesity, hypertension, and type II diabetes.

Some researchers believe that the antioxidant nutrients, including vitamins C and E and vitamin A precursors called *retinoids,* may help prevent some of the age-related changes observed in the eyes, such as SMD. It is possible that these nutrients help buffer the eye from chronic UV damage. Some researchers believe many of the changes that occur as we age are caused by free-radical damage. Antioxidant nutrients may help reduce such damage. (See Essay 3 to review the concept of antioxidants and their dietary sources.)

Medical research suggests that zinc deficiency seems to increase one's risk for SMD. This does not mean that high levels of zinc will either correct or prevent SMD. Zinc is toxic at high doses, so it is best to meet one's recommended daily requirement for zinc by consuming foods supplying this nutrient, such as fish, chicken, red meat, whole grains, and legumes.

ESSAY

17

MEDITATION AND BIOFEEDBACK: ROADS TO THE RELAXATION RESPONSE

Western scientists once assumed that the functioning of the autonomic nervous system (ANS) was outside of voluntary control. In the 1960s interest in Eastern religious practices combined with the growing technology of biofeedback to produce evidence that convinced the medical community that the ANS was not completely "automatic." Yogis were observed to decrease metabolic rate dramatically during meditation. In the laboratory, rats learned to slow their heart rates to receive rewards, and people voluntarily lowered their blood pressure to hear fewer biofeedback beeps.

Meditation

Although new to Western medicine, countless forms of meditation have existed for thousands of years. Most types of meditation consist of attempting to focus one's attention on one thing at a time. The object of focus is often a word or phrase that is repeated over and over, for example, a prayer. Some people meditate on a visual object such as a candle or a flower. A focus on the breathing process is also common, and may be combined with counting or repetition of a phrase. Some meditators simply observe the thoughts that flow through their minds.

As the mind becomes quiet and focused, physical relaxation occurs as well. Herbert Benson has called this state the *relaxation response* and was one of the first Western physicians to bring the medical benefits of meditation to the attention of North Americans.

Biofeedback

Biofeedback instruments provide information about what is happening in your body and enable you to use this information to gain control of the variables being monitored. For example, people who suffer from tension headaches can use biofeedback to become aware of muscle tension and learn to relax. Electrodes are placed over one of the offending muscles, typically the frontalis muscle of the forehead. The electrodes detect muscle electrical activity and send this information to the biofeedback instrument, which converts it into a signal such as a beep or flashing light that can be perceived by the patient. As the patient relaxes the muscle, the beeps or lights will slow down; if the muscle gets more tense, they will speed up. The instrument responds instantaneously to any change in muscle electrical activity, so the patient receives immediate feedback on the success of the relaxation practice.

If a body response can be monitored, biofeedback can be applied. Those most commonly monitored are those most accessible. Electromyography (EMG) senses the electrical activity of muscles, as in the example above. Electroencephalography (EEG) gives information about brain wave activity. Using EEG, subjects learn to produce the types of brain waves associated with relaxed or creative mental states. The galvanic skin response (GSR), also known as electrodermal response (EDR), is what lie detector tests use. Changes in the electrical conductivity of the skin reflect minute changes in sweat gland activity and skin cell membrane permeability, which occur in response to stress.

Biofeedback instruments can also measure skin temperature, which reflects vasodilation (opening) of peripheral blood vessels. The more relaxed a person, the greater the peripheral vasodilation, and the higher the skin temperature. Blood pressure, heart rate and rhythm, and even stomach acid secretion can also be controlled with biofeedback training.

Clinical applications

Meditation and biofeedback are most commonly used to reduce sympathetic arousal and achieve a more "parasympathetic" homeostasis. They are effective for the treatment of many stress-related illnesses such as ulcers, colitis, irritable bowel syndrome, hypertension, tension and migraine headaches, and muscle tension problems such as low back pain. They are also helpful for a disorder called *Raynaud's disease,* whose symptoms are cold extremities due to peripheral vasoconstriction. People with Raynaud's disease can learn to warm their hands and feet by dilating constricted arteries. Temporomandibular joint (TMJ) problems and insomnia are also sometimes treated with meditation and biofeedback. Psychological disorders, such as anxiety and phobias, often respond to meditation and biofeedback training.

Biofeedback is also used for certain forms of muscle rehabilitation. It can help a person regain use of muscles following stroke or trauma. Biofeedback instruments can sense very low levels of muscle activity and help the patient learn to increase the level to produce a muscle contraction. Biofeedback is also used to help reduce the firing of nerves that produce muscle spasms and spastic movement.

How does it work?

No one knows exactly how autonomic functions are brought under voluntary control. Even somatic functions that we call voluntary are not always subject to conscious control. For example, skeletal muscles are governed by the voluntary nervous system, but a skeletal muscle spasm may not stop even though you will it to do so.

People learning to use biofeedback usually work with a therapist who adjusts the instruments and gives suggestions. When biofeedback is used to decrease sympathetic arousal, patients usually practice a relaxation technique such as meditation. If they are trying to warm their hands, they might imagine lying in the sun (with plenty of sunscreen) or putting their hands into very warm water, making such an image the focus of meditation. As the beeps slow down, they begin to associate certain images, thoughts, and sensations with relaxation. Important in this process is an attitude of passive attention, a "trying not to try." Willing the arteries in the hand to open only leads to more tension and less opening. Instead the person must simply be aware, tune in, relax, and notice what seems to work.

An important part of meditation and biofeedback training is learning to transfer the skills learned during practice sessions to real-life situations, especially if the goal of training is daily physiological control. A person must be able to regulate blood pressure while driving in traffic, talking to friends, and performing a job, not just when meditating in a quiet room or hooked up to the biofeedback machine.

As research into voluntary control of the ANS continues, one of the most intriguing observations is that the more deeply into the nervous system you get, the harder it is to tell where the body ends and the mind begins. This illustration of the intimate connection between thoughts and emotions and autonomic and somatic functions supports the notion that a psychosomatic (literarally "mind-body") illness is not "all in a person's mind."

ESSAY

18 STRESS RESISTANCE: IT DEPENDS ON YOUR POINT OF VIEW

The "fight-or-flight" arousal of the sympathetic nervous system is very helpful when you encounter a snarling dog or need to escape from a burning building. But how often in real life do we encounter a source of stress that gives us a chance to fight or run away? Traffic jams, problematic professors, family conflict, financial problems: instead of fight-or-flight, it is sit-and-stew.

Eustress versus distress

Contrary to popular opinion, stress is not always bad, and the physical reactions we have in response to stress are not necessarily harmful to our health. While the word *stress* carries a negative connotation, stress occurs whenever we must adapt in some way to a stimulus. It might be something as simple as a telephone ringing; we "adapt" by recognizing the source of the noise and picking up the phone. Or the stimulus might be something more complicated, like a friend asking to borrow our car. Our thoughts can be major sources of stress.

Many types of stress are *positive;* such stress is called *eustress.* Sources of eustress include holidays, going to a party, giving a presentation, getting married, job promotions, buying a house, vacations, and travel. Sources of eustress are associated with outcomes expected to be mostly positive, and with some sense of personal control over that outcome. On the other hand, sources of *distress* are associated with outcomes expected to be mostly *negative,* and with stress over which we feel little personal control. Expectation and control are the two key elements that distinguish eustress and distress.

This distinction between distress and eustress sounds simple, but as sages through the ages have observed, stress is often caused not so much by what is actually happening, but rather by our perception of what is happening. Of course, some sources of stress are seen as distress by almost everyone, for example, the sickness or death of a loved one, divorce, or loss of a job. But most everyday stressors need not be as distressing as we make them.

Think about a potentially stressful family gathering. Focus on the people

you prefer not to see, their negative comments and opinions, what they might be saying about you, all the work involved in the preparation and clean-up. And now control. Imagine you have no option but to attend the party. You have no control over who will corner you in conversation, and you must behave according to your historical role in the family. Are we having fun yet?

Now focus instead on the people you look forward to seeing, the questions you'd like to ask them, the children who are a year older, the good things to eat, the special family rituals. Pretend you are in your nineties looking back on your life. Consider the control you do have. You are choosing to attend this important event. Decide which people you will spend the most time with, and whom you will avoid. What are some things you could do to make the gathering as pleasant as possible? A walk with your favorite uncle? A shared confidence with a favorite cousin? Whatever you imagine, the secret to the creation of eustress from potential distress lies in finding a meaningful perspective on the upcoming event and devising a plan to change a potentially distressing occasion to one of value for you.

Stress and health

Whether we perceive stressors as eustress or distress has important implications for our health. While at one time it was thought that "too much" of any kind of stress increased a person's risk for stress-related illness, recent evidence suggests that our bodies respond differently to positive and negative stress. In particular, the hormonal changes that occur in response to eustress are less harmful than those produced in response to distress. Under many conditions these physiological responses to eustress help us cope more effectively with the source of stress. Negative health consequences associated with stress, such as decline in immune system function (see Essay 21), are more often associated with distress than eustress.

Early research on Type A personality and heart disease suggested Type A people were more prone to heart attack. Type A people were characterized as time-conscious high achievers who hate to wait in line and are always in a hurry. Subsequent research has shown that it is not the general Type A behavior pattern that is related to heart disease, but rather certain elements associated with this behavior pattern, specifically feelings of anger, hostility, and distrust.

People who view most stress as distress seem to upset the physiological balance of the sympathetic and parasympathetic branches of the autonomic nervous system. Cynical, hostile people spend too much time wound up, high on epinephrine, norepinephrine, and other hormones, such as cortisol, which contribute to hypertension and artery disease.

Can you change your stress response?

Our stress styles develop throughout the course of our lives, partly as learned behavior. They can be unlearned with dedicated practice. Stress-management courses and workshops can help us learn better ways to cope with stress. We can improve our ability to deal directly with sources of stress, learn how to say no, and manage our time more effectively. We can monitor our thoughts, question mistrusting assumptions that may not be true, and replace them with more open and positive observations.

Cultivating a sense of humor and an ability to relax are important. Many stress-management programs teach exercises that promote the relaxation response, including biofeedback and meditation (see Essay 17). Developing better communications skills, learning to listen, and practicing trust and forgiveness can help us cultivate what cardiologist Redford Williams has called a "trusting heart." Experiencing the pleasure of volunteer work that involves helping others can also help heal the heart of the person who tends to see the glass as half empty rather than half full.

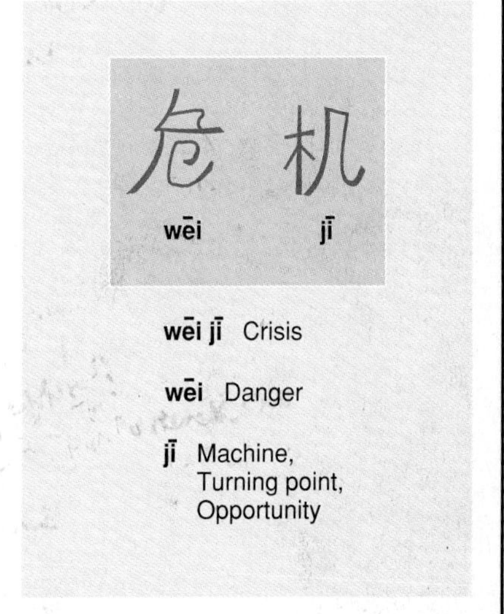

The Chinese character for *crisis* is a good representation of the dual nature of stress, since it is formed from the characters for *danger* and *opportunity.*

ESSAY

19 SAFE TRANSFUSION: FACT OR DELUSION?

The ultimate irony. You've just been through weeks of complicated and painful medical procedures. Fortunately, modern medicine triumphs again: The surgery was successful, and now you're on the road to recovery. But some routine blood work following your hospitalization reveals that you've acquired another potentially fatal disease, probably from blood received during surgery. The very treatment designed to solve one problem has created another.

How safe is the blood supply? This question was rarely asked before AIDS (acquired immune deficiency syndrome) became a household word in the mid-1980s. At this time health officials realized that HIV (human immunodeficiency virus), the virus that causes AIDS, might be present in the blood of some donors. And indeed, some people who received transfusions before blood could be tested for HIV antibodies did acquire the virus. Everyone began to ask, could this happen to me?

We've come a long way

The concept of blood transfusion dates back to the seventeenth century, when animal blood was experimentally transfused into humans. (It didn't work.) In the 1800s blood transfusions with human donors were attempted with a little more success. But blood transfusions did not become a useful medical procedure until the understanding of blood-type compatibility in the early 1900s and the discovery of sodium citrate to slow clotting in 1914. The development of cold storage of blood in 1940 allowed blood to be stored for up to three weeks, thus increasing the feasibility of banking blood. In the U.S., organized "blood drives" to acquire blood for use in transfusions first began in 1947.

Blood transfusion is important for many medical procedures. About 17% of adults in the U.S. have had a blood transfusion at some time, and about 2% of the population per year receive them. The fact that blood transfusions have become quite commonplace is one of the reasons for the dramatic rise in concern over the safety of the blood supply: A blood transfusion could be in anyone's future.

What are the risks?

Homologous blood transfusions (in which blood is received from a donor other than oneself) have always carried some risk of acquiring an infectious disease. This risk has usually been considerably outweighed by the benefits of the medical procedure. And fortunately, screening for blood-borne viruses has improved considerably in recent years. Most medical researchers agree that a "zero-risk" blood supply is an impossibility. We can reduce the risk as much as possible and weigh existing risks against benefits when deciding whether to perform a blood transfusion.

How much risk of acquiring HIV from a transfusion are we willing to accept? Since AIDS is a fatal disease, even a small risk remains significant in the eyes of most people. The biggest concern arises over our inability to detect the presence of HIV during a potential donor's first three months of infection. To combat the possibility of blood donation during this period, blood collection agencies attempt to discourage blood donation from people whose behavior puts them at risk for contracting the virus.

Statistics show that the risk of transfusion-acquired HIV infection is currently estimated to be about one in 153,000. Compare this to an estimated risk of 1 in 6,000 for hemolytic transfusion reaction (in which red blood cells rupture) and a risk of 1 in 100,00 for fatal hemolytic transfusion reaction.

The most common serious complication associated with homologous blood transfusion is hepatitis infection. (*Hepatitis* refers to inflammation of the liver, which, in the case of transfusions, may be caused by several viruses.) In 1989, risk for hepatitis B infection was 1 in 200 to 1 in 300, and risk for hepatitis C infection was estimated to be 1 in 100. Screening procedures first became available for the hepatitis C virus in 1990, and risk for this infection has declined significantly since then.

Donated blood is also screened for other viruses, including syphilis and adult T-lymphocyte virus type 1 (HTLV-1). This rare virus (present in 0.025% of all U.S. blood donors) is associated with a certain type of leukemia.

Alternatives to homologous blood donation

Despite the improved safety of the blood supply, alternatives are still actively sought by many people. One such alternative is *directed donations.* This is when specific people are recruited to donate blood for a given patient. At first glance this might appear to reduce infection risk, but recent data show that blood obtained in this fashion is no safer than blood donated by random donors. One explanation for this is the possibility that potential donors who might be at risk for carrying a pathogen would feel pressured to give blood, instead of being discouraged by the prescreening questionnaire designed to eliminate blood donation from people with risky behaviors. Since confidentiality could not be ensured for directed donations, friends of the patient might have more difficulty refusing requests to donate blood.

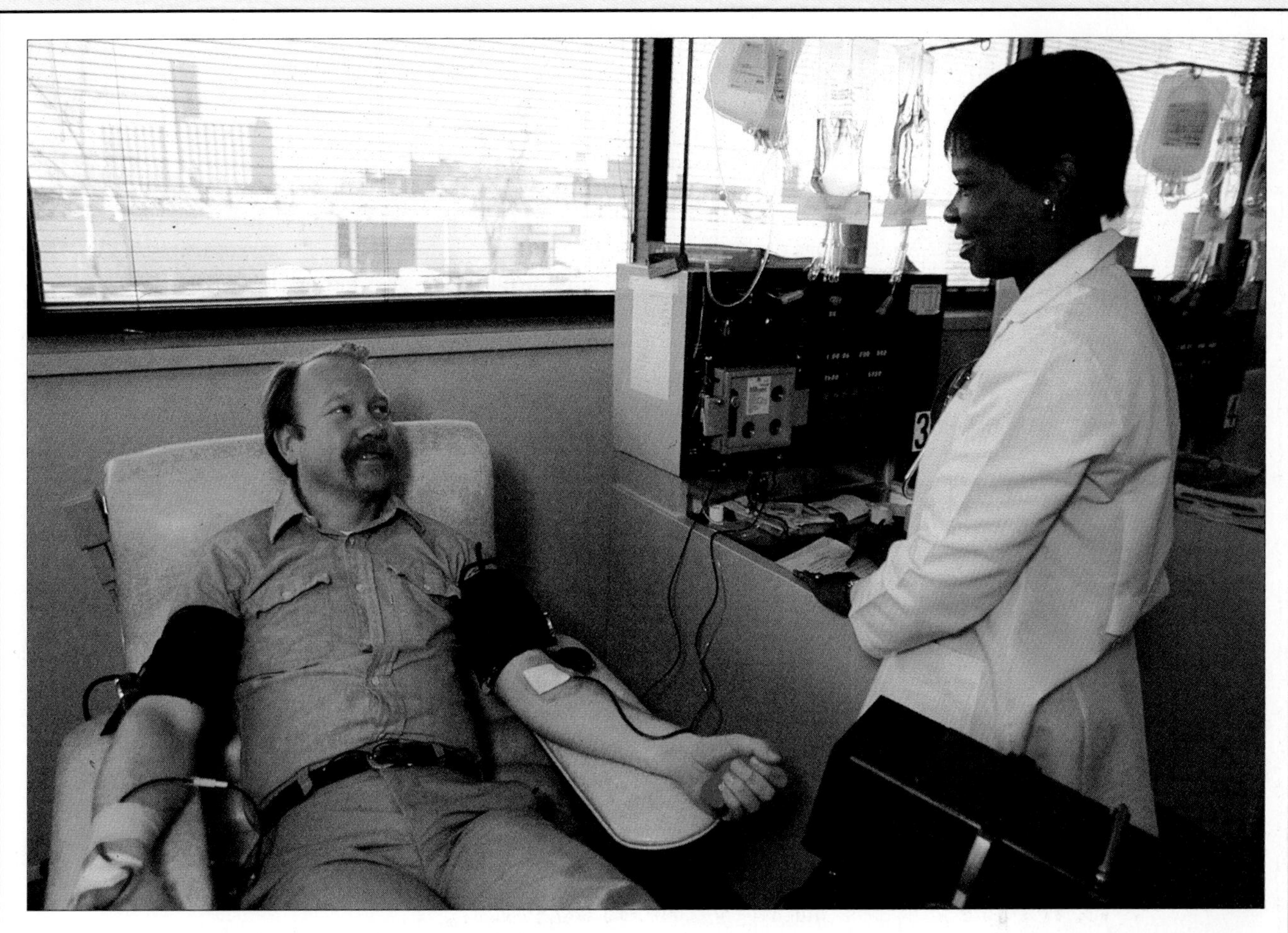

Autologous donation is the donation of one's own blood for later use. Autologous donation increased 15-fold in the 1980s. Autologous donation is most often done by patients with upcoming elective surgery. Predonation ensures an abundant supply of blood, reduces transfusion risks such as becoming infected with AIDS or hepatitis viruses, and avoids a fatal transfusion reaction. Autologous donation also eases the demand on blood banks and provides an alternative for people who object to receiving banked blood for certain religious reasons.

Autologous intraoperative transfusion is a procedure in which blood lost during surgery is suctioned from the patient, treated with an anticoagulant, filtered of debris, and centrifuged to recover the red blood cells. The red blood cells are then washed in saline solution and reinfused.

What about simply putting some of your own blood away for a rainy day? Storing one's own blood for some potential unforeseen emergency is not currently endorsed by health officials for several reasons. First, the procedure is costly. Second, since blood products have a limited shelf life, the procedure would need to be repeated quite often. And third, those in charge of the blood banks feel unable to guarantee that the blood could be shipped rapidly enough in the event of emergency.

ESSAY

20 DIET AND HEART DISEASE: REDEFINING THE GOOD LIFE

A person's diet has a strong influence on four risk factors associated with heart disease: high blood cholesterol, high blood pressure, obesity, and diabetes mellitus. So people watching their cholesterol, blood pressure, weight, or blood sugar should also be watching their diets.

Public health officials have issued a number of dietary guidelines and recommendations to help people eat a "heart-healthy" diet, a diet that can help prevent coronary artery disease, the leading cause of death in North America. There's no one food that can prevent heart disease, or even a short-term eating plan that will reverse atherosclerosis. A heart-healthy diet consists of making the right daily food choices that add up to a lifetime of good dietary habits.

Eat less fat, saturated fat, and cholesterol

We need only a very small amount of fat in the diet to maintain good health; extra fat is extra calories, which contribute to obesity. Recent research shows that a given number of calories consumed as fat results in more fat storage than the same number of calories consumed as carbohydrate or protein. In other words, it is metabolically efficient to store fat in adipose tissue, but not as efficient to turn carbohydrate or protein into fat.

A diet high in fat, particularly saturated fat, also raises blood cholesterol level. Foods high in saturated fat include butter, cream, whole milk, cheese, some shortenings and margarines, and palm and coconut oils. Some food labels list how much and what types of fat have been added to food products. Lean cuts of meat, poultry, and fish have less saturated fat than fatty cuts such as prime rib. Chicken fat is found under the skin, so by trimming the skin you remove most of the fat. Cooking methods such as broiling and baking are preferable to frying.

Cholesterol intake should be limited by restricting consumption of egg yolks, meats, and organ meats. Cholesterol is found in muscle tissue itself, so simply trimming the visible fat from meats does little to decrease its cholesterol content.

Eat more fish

Research has suggested that people who include fish in their diet have a lower risk of heart disease. Cold-water fish such as salmon, mackerel, trout, bluefish, and herring contain omega-3 fatty acids. These fatty acids may prevent heart disease by decreasing the tendency of platelets to clump and perhaps by lowering total serum cholesterol as well.

People who consume diets high in fish oils produce less thromboxane, which promotes platelet clumping and is a powerful vasoconstrictor. Fewer clots and more open arteries mean a lower risk of blockage. Although fish oil contains substantial amounts of cholesterol, some studies have shown that people who consumed fish oil reduced their serum cholesterol levels.

Scientists are reluctant to recommend fish oil supplements at this point, because their safety for long-term consumption is not known. But eating more fish is a good idea. A study from the Netherlands found that people who eat only one or two fish meals a week, regardless of the type of fish consumed, had a lower incidence of heart disease than people who ate less fish.

Eat more foods high in complex carbohydrates and fiber

Carbohydrate foods do not increase total serum cholesterol and are less likely to cause obesity than fatty foods, although too much of any food means extra calories. Complex carbohydrates are especially good because they have a lot of nutrition per calorie and are rich in vitamins and minerals. They are also high in the type of fiber water-soluble) that helps decrease serum cholesterol. Complex carbohydrates include grains such as wheat, rice, corn, oats, and their products such as cereals, breads, and pasta; peas and beans, such as split peas, lentils, kidney beans, and chick-peas; and starchy vegetables such as potatoes, yams, and winter squashes.

Some research has suggested that people consuming a diet high in complex carbohydrates can maintain or even lose weight while eating as much as they want to of these foods. They don't count calories or portions, and eat until they are full. Researchers theorize that people consuming a diet high in these foods feel full and quit eating before they have consumed too many calories. Just the opposite can happen with foods high in fat. Since a lot of calories are contained in a small volume, you can eat hundreds of calories in a very short period of time.

It is unfortunate that dieters have shunned complex carbohydrates for years. Many people would still omit the potato rather than the steak when trying to lose weight. Yet even a small (3.5 oz) steak is high in fat and may have over 400 calories, while a baked potato has almost no fat and only about 90 calories. It's important to prepare these complex carbohydrate foods without high-fat sauces or other added fats, however. The 90-calorie baked potato becomes a high-fat, high-calorie dish as soon as you add butter or sour cream.

Eat less sugar and salt

Simple sugars are found naturally in fruits, vegetables, and dairy products. Refined sugars (sucrose, dextrose, syrups, and other sugar products) provide empty calories; that is, they have no other beneficial nutrients. Moreover, they are often consumed with fats in pies, pastries, cakes, cookies, and candies, all high-calorie foods.

Hypertension (high blood pressure) is more prevalent in population groups that consume high levels of sodium. A taste for salt (sodium chloride) developed early in life can lead to hypertension in later years. Some research suggests that it may take years for a high salt intake to cause hypertension, so most people do not realize that their salt intake is harming their health.

Sodium in the form of salt or monosodium glutamate (a "flavor enhancer") is found in most prepared foods, such as soups, sauces, and canned fish and meat; condiments such as soy sauce and steak sauce; pickled and cured foods; and salty foods such as potato chips and pretzels. High sodium levels are also present in some vegetables, such as celery and mushrooms.

Drink less alcohol

Too much alcohol is associated with increased risk of hypertension. No one can say exactly how much is too much, but most authorities agree one or two drinks a day are probably safe. Alcoholic drinks are empty calories and can also contribute to obesity.

ESSAY

21 HYPERTENSION PREVENTION

Despite the prevalence of hypertension (high blood pressure), a cure still eludes medical researchers. Hypertension medication can be lifesaving, but it does not correct the underlying cause of the disorder. Indeed, for most cases of hypertension the cause is not known. At one time, it was thought that a rise in blood pressure was a normal and necessary part of the aging process, which is why hypertension not caused by some other underlying disease was called *essential hypertension*. Increased blood pressure was thought to be essential to maintain adequate circulation in older adults.

But blood pressure does not increase with age in many other cultures, especially in nonindustrialized countries. Yet 50% of all Americans have hypertension by age 74. What is it about the American way of life that makes us so prone to hypertension? Can an understanding of the relationship between life-style and the development of hypertension help us prevent this disorder?

Risk factors

Hypertension appears to have both genetic and environmental causes. Many risk factors for the development of hypertension have been identified. Unmodifiable risk factors include family history of hypertension (people whose parents or siblings developed hypertension are at increased risk), age (risk increases with age), ethnicity (black Americans have greater risk), and gender (men have more hypertension between the ages of 18 and 54, and women have more hypertension after age 65).

People who fall into a high-risk group should not worry. (Worrying will only increase your blood pressure.) A wellness life-style can significantly reduce a person's risk of developing hypertension and decrease its severity should it develop. Life-style change is particularly effective in helping to lower blood pressure for those with borderline or mild hypertension, 80% of hypertensives.

Diet

As described in Essay 20, a healthful diet can help prevent hypertension. Almost everyone has heard that minimizing sodium intake is important. Sodium sensitivity often develops with age, so even though excess sodium does not currently raise your blood pressure, it may in the future. Anyone at risk for hypertension, and indeed anyone planning to grow old, may benefit from not acquiring a taste for salty food. Animal studies also suggest that salty diets may damage arteries even without causing an increase in blood pressure. In one study, rats fed high-salt diets experienced only a slight rise in blood pressure, but all died much earlier than littermates on a low-salt diet. Autopsies revealed a great deal of arterial damage, including atherosclerosis, and associated damage to brain cells supplied by the damaged arteries. Epidemiological studies support the dangers of a high-salt diet as well. For example, people who live in northern Japan consume high-salt diets and also have a very high incidence of stroke.

Some studies have found that an increased intake of calcium and potassium helps lower blood pressure. Calcium is found in dairy products, dark green vegetables, and calcium-enriched products such as some brands of orange juice and soy milk. Potassium is found in many fruits and vegetables.

An adequate intake of magnesium may also help regulate blood pressure. In animal studies, magnesium deficiency increases vascular muscle tone, thus increasing peripheral resistance to blood flow. Magnesium is found in legumes, nuts, whole grains, and meat.

Alcohol consumption has been strongly associated with hypertension: The greater the alcohol intake, the greater one's risk of hypertension. Many people find that their blood pressure improves when they decrease their alcohol consumption. Caffeine is a stimulant and increases blood pressure, so consumption should be limited if one has or is at risk for developing hypertension.

Obesity

People who are more than 20% above their desirable weight have twice as much hypertension. Extra fat in the abdominal region is more strongly associated with hypertension than excess fat in the hips and thighs. A small weight loss can significantly decrease blood pressure, even if the person does not lose all of the extra weight.

Abdominal obesity is also associated with insulin resistance. People with insulin resistance produce enough insulin, but the insulin receptors located in the cell membrane are not sensitive enough to respond correctly, and blood insulin levels rise. Insulin stimulates sympathetic nervous system activity, similar to the stress response that causes blood pressure to rise. Insulin may also lead to fluid retention, which increases blood pressure. Weight loss often helps restore insulin sensitivity.

Smoking

Nicotine is a potent vasoconstrictor, so here's one more reason not to smoke. When a cigarette is smoked, nicotine enters the blood immediately, and vasoconstriction occurs systemically. The more nicotine and the more frequently one smokes, the greater the vasoconstriction. Smokers usually have chronically elevated blood nicotine levels and increased vascular resistance.

Physical activity

Many studies have found that regular aerobic exercise can normalize blood pressure in people with borderline hypertension. Epidemiological studies have also found that exercise can improve the health of people with this condition. In fact, one study showed that hypertensives who exercised regularly had almost the same mortality rates as people with normal blood pressure.

Exercise may exert this protective effect partly by contributing to a person's weight control efforts. Regular aerobic exercise may also lead to decreased vascular resistance, perhaps by increasing parasympathetic output (the relaxation response) when the person is at rest.

Insulin sensitivity improves for many hours after exercise. As mentioned above, this could prevent the sympathetic nervous system activity and fluid retention that could contribute to high blood pressure.

Many people find that exercise helps them relax and decreases their stress reactivity. Things don't upset them as easily. Several studies have demonstrated an association between stress and hypertension, so exercise may be helpful because it reduces a person's stress response.

Stress management

It has been hypothesized that chronic stress may lead to a chronic elevation in sympathetic nervous system activity. This includes an elevation in the hormones associated with the stress response, including epinephrine and norepinephrine, which might cause chronic high blood pressure. Stress-management techniques can help a person learn relaxation skills that decrease sympathetic output and thus reduce blood pressure. (For more information on relaxation and stress management, see Essays 17 and 18.)

ESSAY

MIND AND IMMUNITY

22

Subjects in a study designed to evaluate the effectiveness of a painkiller experience symptom relief after receiving inactive "sugar pills." People with asthma suffer asthma attacks when they believe they are being exposed to allergens, even though the allergens are purely imaginary. Students experiencing the most psychological stress during exam period come down with more colds than their easygoing friends.

Researchers are beginning to explore the physiological bases of the body–mind relationship and to elucidate the ways our thoughts and feelings can influence the physiological processes that affect our health. Many scientists are discovering that we can best understand physical health with a holistic outlook that includes the functioning of the mind. Psychoneuroimmunology (PNI) is the study of the interrelationships of the three body–mind systems that serve as communication networks in the orchestration of homeostasis: the nervous, endocrine, and immune systems.

Making the connection

At one time physiologists thought the immune system functioned independently. Now, anatomical and physiological evidence points to extensive communication between the immune system and the nervous and endocrine systems. Several organs of the immune system, including the spleen, thymus, and lymph nodes, receive sympathetic nervous system (SNS) innervation. In animal studies, when this innervation is tampered with, changes in immune responses occur. In the spleen and thymus, the SNS nerve endings have a synapse-like connection (the type of connection observed in the nervous system) with lymphocytes (a type of white blood cell), and appear to communicate in some way with these immune cells.

In experiments with laboratory animals, injury to certain areas of the brain leads to immune system changes. These changes affect *specific resistance to disease,* which refers to the immune system's ability to recognize, attack, and remember particular foreign molecules such as viruses and bacteria. Specific resistance to disease is carried out by several types of white blood cells called *B cells* and *T cells.* Lesions to one part of the brain (the anterior hypothalamus) interfere with the activity of antibody-secreting B cells, while damage to another area (the posterior hypothalamus) reduces the effectiveness of the memory T cells.

Neurotransmitters, a type of chemical messenger found in the nervous system, are also found in the immune system, particularly in the bone marrow and thymus gland. Many types of immune cells have receptors for neurotransmitters and hormones, and it has been demonstrated that the immune system can respond to neurotransmitters and hormones. For example, at low concentrations, norepinephrine stimulates the immune system. Cortisol, a hormone secreted by the adrenal cortex in association with the stress response, acts to inhibit immune system activity, perhaps one of its energy-conservation effects. An elevation in cortisol has been observed in a significant number of people with major depressive illness, so endocrine effects may help explain the link between depression and decreased immune response.

Some researchers believe the opioid peptides (such as endorphins, discussed in Essay 14) may be another vehicle for communication. Opioid peptides are manufactured by some lymphocytes as well as by nerves in the central nervous system (CNS), which comprises the brain and spinal cord. Opioid receptor sites have been found on some white blood cells.

Communication between the immune system and the CNS goes both ways. The immune system does not merely respond to messages from above, but seems to inform the brain and endocrine organs. Small protein hormones called *cytokines* are secreted by many types of white blood cells and also in the brain. Cytokines appear to act as chemical messengers in this communication network.

The whole is greater than the sum of its parts

Such physiological data are intriguing, particularly in light of studies on humans investigating changes in immune function following various psychosocial stressors. Most readers are familiar with stories of lowered resistance to disease in people experiencing death of a loved one, divorce, job loss, depression, loneliness, exams, and sleep deprivation. But finding the psychophysiological explanation for these observations still eludes us. The challenge to scientists is to figure out exactly how PNI can explain real-life changes in immune response. Humans are especially difficult to study because when under stress, behavioral variables are apt to change as well: People may get less sleep, eat differently, drink more alcohol, take drugs, and so forth, all of which may influence immune responses.

Despite these obstacles, PNI researchers are making some headway in understanding the types of effects our thoughts and feelings can have on the immune system. One study found that men who exhibited a strong SNS response to frustration showed a greater number of suppressor T cells, which inhibit immune response, and a lower rate of lymphocyte proliferation than men with a milder response. Another study found no relationship

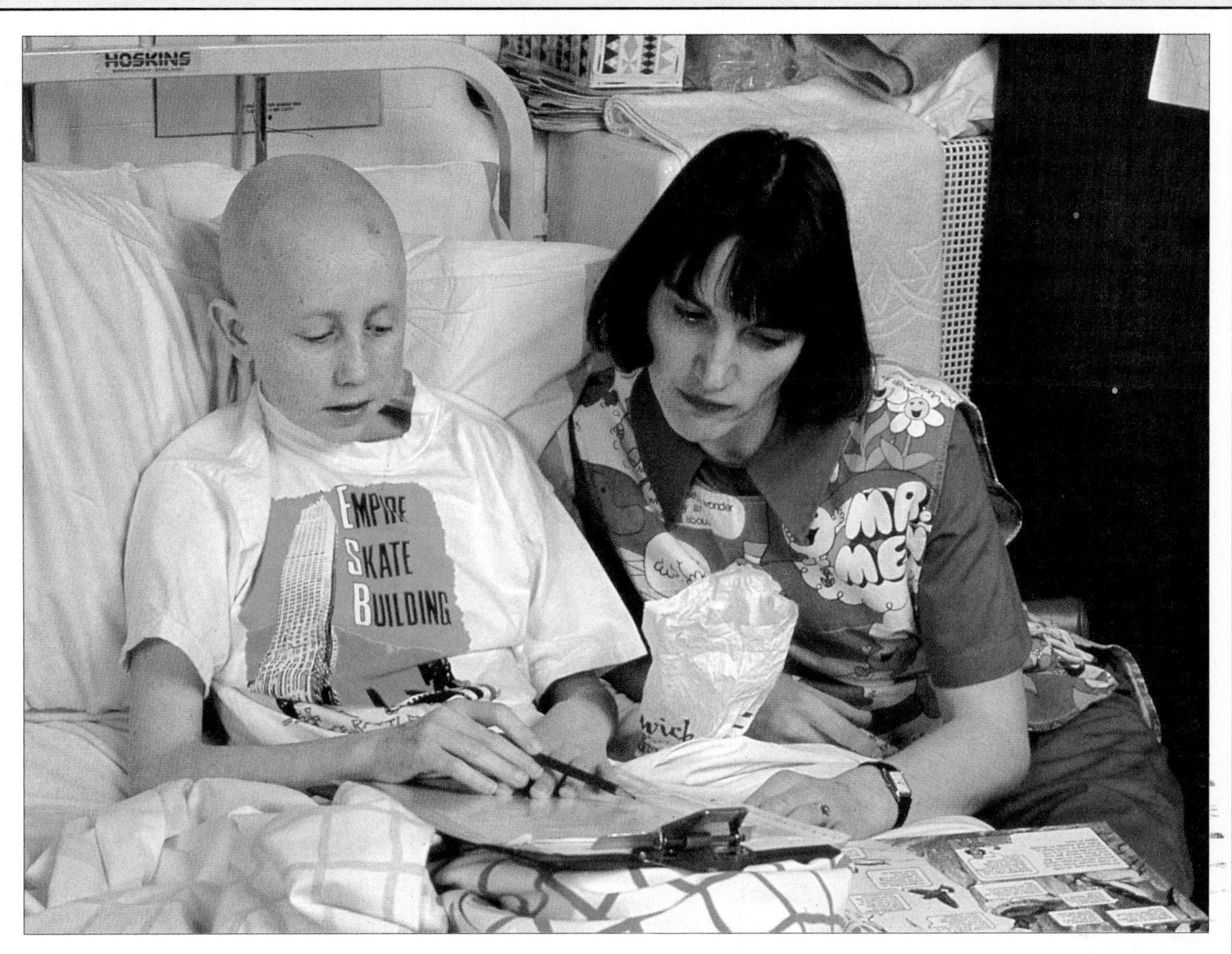

between stress and numbers of lymphocytes or other immune cells, but did find certain changes in lymphocyte proliferation and antibody reactions. Stay tuned!

Clinical applications

PNI research appears to justify what many health-care workers have observed since the beginning of time: One's thoughts, feelings, moods, and beliefs influence one's level of health and the course of disease. Especially harmful appear to be feelings of helplessness, hopelessness, fear, and social isolation, all common to the experience of being sick, and especially to being a patient in a hospital. While conclusive studies are still lacking, preliminary data suggest that the use of positive imagery, relaxation techniques, psychotherapy, and humor can help patients feel much better, and may have a positive influence on the course of a disease.

ESSAY

23 GIVING VIRUSES A COLD RECEPTION

The respiratory system bears the brunt of the suffering when we catch a cold (or in reality, when a cold catches us!). Cold symptoms are not produced directly by the cold virus, but by the body's nonspecific defenses as they fight the virus.

The immune system learns to recognize specific disease agents through exposure to them. This exposure can occur through vaccination or by natural means. We acquire immunity to chicken pox, measles, mumps, tetanus, cholera, smallpox, and many other life-threatening diseases. So, one might wonder, if the immune system can do such amazing things, why can't it defend us from the common cold?

We are susceptible to at least 200 different cold-causing viruses. The most common types, rhinoviruses (literally "nose viruses"), cause about 40% of all colds in adults. As soon as the immune system learns to recognize and defend us from one, another comes along, and then another. This viral diversity creates quite a challenge for the immune system, so much so that most people succumb to one to six colds per year.

Cold prevention

Research evidence indicates that cold viruses may be transmitted from the hands of an infected person to the hands of a susceptible person. The virus can survive on the skin for only a few hours and must reach the nose in order to invade the body. On the face near the nose is no good, because the skin provides an effective barrier. Since the mucous membranes of the mouth are also an inhospitable environment, kissing seldom spreads colds.

If all goes well for the virus, eventually the hand delivers the virus to its new home, the person's respiratory system, by touching the mucous membranes of the nose or the eyes (the virus can travel down the tear duct to the upper nose). One study found that 40 to 90% of people with colds had rhinoviruses on their hands. The viruses were also found on about 15% of nearby objects, such as doorknobs, telephones, and coffee cups. Some

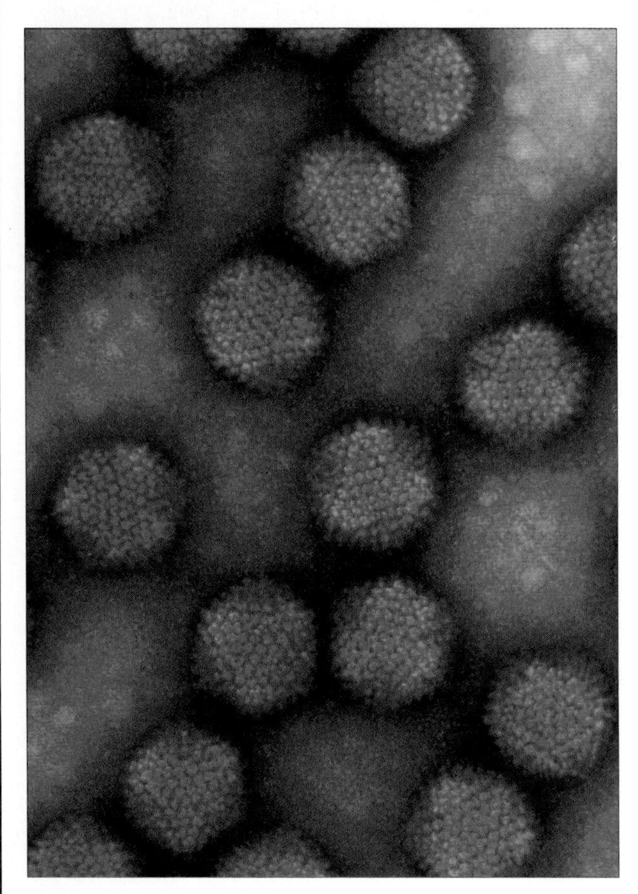

Cold Viruses

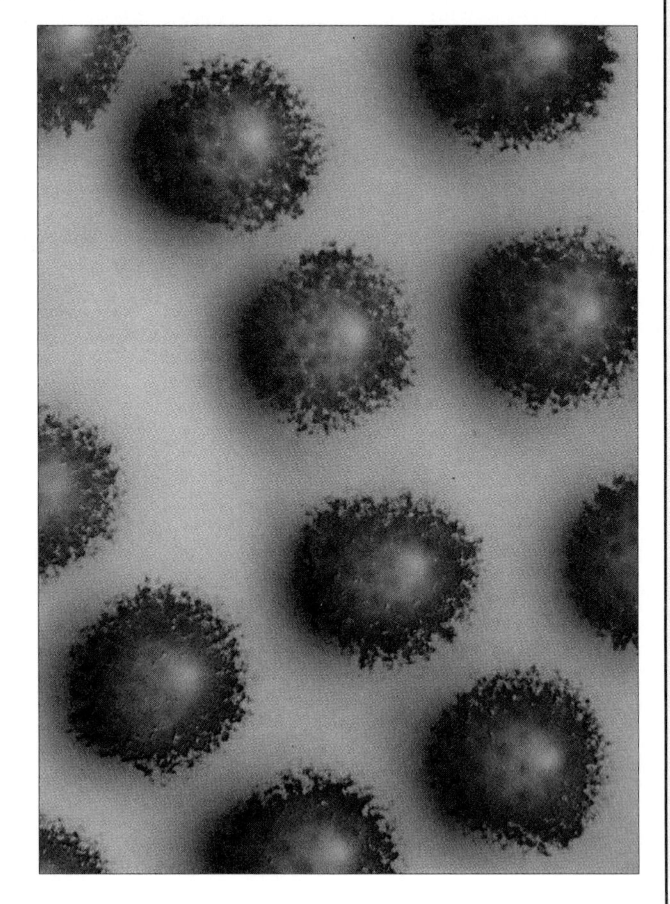

Cold Viruses

evidence also suggests that cold viruses can fly through the air with the greatest of ease and find new homes in new noses. They may spread when expelled by coughing and sneezing.

It is not known what makes some people more susceptible to colds. Small children are the most susceptible, because their immune systems are still immature and haven't learned to recognize as many pathogens. People who are around children a lot also get colds more often. Smokers are more likely to catch colds than nonsmokers, partly because smoking inhibits the airway cilia that help move mucus. Some studies have shown that stress can decrease the effectiveness of the immune system, and some evidence suggests that stress and fatigue increase susceptibility to colds (see Essay 22).

Given what we know about the transmission of colds, the single best way to prevent colds is frequent handwashing, especially when you're around people who have colds. Avoid sharing telephones, glasses, towels, and other objects with a person who has a cold. And try not to touch your nose or eyes. People with colds should cough and sneeze into facial tissues and then throw the tissues away.

Getting enough rest, eating well, exercising moderately, and managing stress certainly won't hurt and may help keep your resistance up. If you're a smoker, cold prevention is yet another good reason to quit.

What about vitamin C? Although studies have failed to show that vitamin C prevents colds, some research has found that it may lessen the severity of cold symptoms. Vitamin C also increases the integrity of cell membranes and so may make them harder for viruses to penetrate.

Cold self-care

Since a cure for colds continues to elude medical researchers, the best we can do is to treat the symptoms. It's been said that with aggressive medical treatment a cold will disappear in seven days, while if left alone a cold will last a week. Nevertheless, treatment of symptoms can at least make us feel better until the cold has run its course.

The first step in cold self-care is to decide whether your symptoms are those of a cold or something more serious requiring medical attention. People who have heart disease, emphysema, diabetes, or other health conditions should get professional advice before initiating self-care, especially before taking over-the-counter medication. Pregnant and lactating women should also check with their doctors before taking any medication.

Symptoms that indicate your infection may be more than a cold include:

1. Oral temperature over 103°F (39.5°C).
2. Sore throat with temperature above 101°F (38.5°C) for over 24 hours.
3. Temperature over 100°F (37.5°C) for three days.
4. Severe pain in ears, head, chest, or stomach.
5. Symptoms that persist more than a week.
6. Enlarged lymph nodes.
7. In a child, difficulty breathing, or greater than normal irritability or lethargy.

Once you decide you have a cold, there are several things you can do to help yourself feel better. These include:

1. Chicken soup, broth, or other hot drinks. Your mother was right: Hot fluids help relieve congestion by increasing the flow of nasal secretions. They also soothe irritated throats.
2. Gargle with salt water (1/4 teaspoon salt in 8 oz water) to soothe a sore throat.
3. Use a vaporizer or humidifier to increase humidity, especially if the air is very dry. Humid air is gentler on nose and throat.
4. Breathing steam gives your nose a temporary fever, creating an inhospitable environment for the virus. It also helps to thin the mucus causing a stuffy nose, and thus temporarily relieves congestion. The steam may also feel soothing to irritated throats and nasal passages.
5. While rest may not hasten your recovery, it may help you feel better. It's good to stay out of circulation for the first few days of a cold to keep others from getting it and to be sure that what you have is a cold and not something more serious.
6. Many over-the-counter cold medications are available. If you decide you need something, avoid combination drugs that contain several active ingredients to treat several symptoms. If, instead, you buy single drugs for the symptoms you wish to treat, you will avoid taking unnecessary drugs and decrease unpleasant side effects.

ESSAY

24 ALCOHOL: USE OR ABUSE?

What scenes come to mind when you hear the words *drug abuse*? Junkies shooting up heroin? Kids smoking crack? What about good old uncle Jim who drinks too much? After all, alcohol is the number one drug abuse and mental health problem in North America. Far more people abuse alcohol (or should we say, abuse themselves with alcohol) than any other drug. Over 20 million Americans consume excessive amounts of alcohol (14 or more drinks a week). Fifty percent of all alcohol is consumed by 10% of all drinkers.

Physiological effects

The digestive system is affected in several ways by overconsumption of alcohol. Alcohol causes an increase in stomach secretions, which can lead to gastritis, or inflammation of the stomach lining. Alcohol also interferes with the ability of the small intestine to absorb and transport nutrients, especially the vitamins thiamine and folic acid, and minerals. Impaired nutrient absorption and transport enhances the tendency of alcoholics to be malnourished. People who consume large amounts of alcohol also tend to have a poor diet. This malnutrition occurs even though the alcoholic is consuming plenty of calories. Alcohol is calorically dense: 7 kcal/g, compared to 4 kcal/g for protein and carbohydrate and 9 kcal/g for fats. Moreover, the calories in alcohol are empty, because they do not supply the protein, vitamins, fiber, or minerals of the foods they are replacing. Heavy drinking can also cause diarrhea, which decreases nutrient absorption from the small intestine.

The organ that sustains the most damage from alcohol abuse is the liver, which processes 95% of the alcohol intake. The liver converts alcohol into acetaldehyde, which is even more toxic to the body than alcohol. This process is associated with the deposition of fatty compounds in the liver, causing fatty liver, a common condition in alcoholics. If drinking stops, this condition is reversible. If drinking continues, it progresses to an inflammatory condition called *alcoholic hepatitis,* and then to *cirrhosis.* In cirrhosis, functional liver cells are replaced by nonfunctional scar tissue, liver function declines, and the result may be death. Damaged livers are also more susceptible to hemorrhage, and risk of cancer of the liver is 30% higher in people with cirrhosis.

Alcohol consumption has been associated with hypertension, which is reversible if drinking stops. Alcohol abuse can also lead to the development of fat deposits in the heart muscle, which weaken contractile force and reduce cardiac output. Alcoholics have an increased risk of cardiac arrythmias and stroke. Muscle atrophy and weakness often result from alcohol abuse. Skin disorders are also common symptoms of alcohol abuse, and result from liver damage, poor nutrition, and the effects of alcohol and acetaldehyde.

Alcohol is a nervous system depressant and impairs brain function. Long-term effects include brain damage, especially in the cerebral cortex. Alcoholics may experience memory loss, confusion, and even hallucinations and psychotic behavior. Damage to the peripheral nerves, called *alcoholic neuropathy,* may also occur.

The immune system may become suppressed if too much alcohol is consumed. Alcohol inhibits the bone marrow's ability to produce the white blood cells that destroy harmful bacteria. People whose immune systems are already weakened due to other illness should be especially careful about alcohol consumption.

Alcohol use has been associated with several types of cancers, especially cancers of the stomach, liver, lung, pancreas, colon, and tongue. Alcohol and smoking together increase the risk for cancers of the mouth, larynx, and esophagus. Some evidence suggests a relationship between alcohol and the risk of breast cancer.

A woman who consumes alcohol while pregnant exposes the developing fetus to alcohol's damaging effects. Drinking while pregnant is one of the leading causes of birth defects in North America. Alcohol diffuses freely across the placenta, so the alcohol level in the fetus's bloodstream is the same as that in the mother's. Alcohol affects the fetus's rapidly growing tissues, especially the brain. Fetal alcohol syndrome (FAS) is characterized by stunted growth, mental retardation, malformed facial features, and heart defects. The damage is irreversible. FAS is found in babies of light drinkers as well as heavier abusers.

How much is too much?

It is difficult to say how much alcohol is too much. Individual tolerance varies greatly, so that two people consuming similar amounts may experience very different health effects. On the average, men who drink more than three drinks a day and women who drink more than one and a half drinks a day experience a higher incidence of liver disease. (A "drink" is equivalent to a 12-oz bottle of beer, a 5-oz glass of table wine, or a cocktail with 1.5 oz of liquor. All contain the same amount of alcohol, about 0.6 oz.) Rates of liver disease rise in proportion to the amount of alcohol consumed. People prone to a drinking problem may find that no level of alcohol consumption is safe if they are unable to stop after the first drink. (See Essay 12 for more information on addiction.)

Women appear to be more susceptible to the effects of alcohol. This is due in part to differences in size. Women are generally smaller than men, and a given amount of alcohol is more concentrated in a smaller body. But there's more to it than size. When men and women of the same size ingest the same amount of alcohol, women tend to have higher blood alcohol levels. Recent research has found that women do not digest alcohol as effectively as men do. The enzyme alcohol dehydrogenase begins to digest alcohol in the stomach. The enzymes of the women in one study broke down less than one-fourth as much alcohol as those of the men. Alcoholic subjects had an even lower stomach enzyme activity. This research helps to explain why women develop liver disease at lower alcohol intakes than do men.

Recommendations for safer drinking

Alcohol is a potent drug, and must be recognized as such. People who choose to consume alcohol should understand alcohol's effects and use it only in a safe environment. People with a family history of alcoholism should be aware of their parents' patterns of alcohol use and develop more effective drinking strategies. The safe drinker drinks for positive reasons: to celebrate, share, and communicate, not to relieve pain, forget problems, or overcome fears. Abstaining periodically from alcohol use helps one avoid developing a tolerance to this addictive drug.

Normal Liver

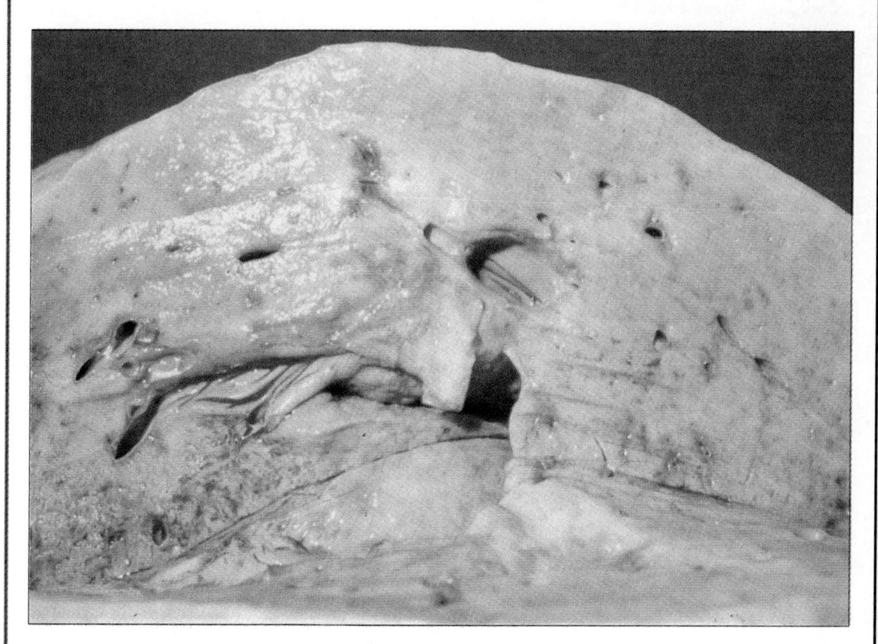

Cirrhotic Liver

ESSAY 25

THE METABOLIC REALITIES OF WEIGHT CONTROL

Losing weight is theoretically simple, a question of metabolic balance. Just adjust activity level and food intake so that the number of calories expended exceeds the number of calories consumed. An energy deficit means that energy stores (one hopes, fat cells) will be used up and not replenished.

Fat is fattening

Calculating the effect of a given caloric intake is not as simple as it sounds. Research suggests that some calories are more fattening than others. It is still true that 1 g of fat contains 9 kcal and 1 g of protein or carbohydrate contains 4 kcal. But a certain amount of energy is required to convert energy nutrients into storage fuels. For example, it is metabolically very efficient to convert triglyceride molecules from dietary fat into triglyceride molecules in fat cells; the conversion is relatively easy and requires little energy. On the other hand, the conversion of dietary carbohydrate into adipose triglyceride is metabolically "costly"; it takes several steps and some energy is actually expended during this conversion process. Fats are more fattening, so a person trying to lose weight will benefit by selecting foods low in fat.

Take the crash out of dieting

Very low-calorie diets, commonly and appropriately known as *crash diets,* are not only ineffective weight loss methods; they are harmful to one's health. These diets typically allow the dieter to consume a very limited and often bizarre selection of foods or food substitutes (usually purchased at a substantial price). Diets such as these do nothing to educate the dieter about prudent food choices, so once the diet is over, the dieter usually gains back all of the weight lost on the diet, and then some.

Common side effect of very low-calorie diets are feelings of deprivation and depression and extremely powerful food cravings. These cravings can lead to food binges, during which the dieter wolfs down mountains of forbidden foods. The binging is usually followed by even stricter dieting, greater cravings, and more misery. It is not uncommon for this harmful pattern to evolve into full-fledged eating disorders.

Much of the recent interest in caloric restriction and metabolic rate is a result of the desire to understand why some people have a great deal of difficulty losing weight, even though they may be eating very little food. According to standard metabolic calculations, they should be losing steadily. But they aren't. It used to be that if you were one of these people who dieted but didn't lose weight, you were suspected of cheating or lying. You were accused of underestimating your food intake, fudging your calorie counts, and failing to follow orders.

But many of the people who do not lose weight on very low-calorie diets may be experiencing a metabolic adaptation to what their bodies perceive as starvation. In the face of starvation, basal metobolic rate (BMR) may decline as the body attempts to conserve energy. This response appears to become stronger the more frequently a person undergoes severe caloric restriction.

Frequent dieting interspersed with periods of normal or greater-than-normal caloric intake are common in people attempting to lose weight. The same 10 or 20 pounds may be lost and regained many times, in what has become known as the rhythm method of girth control, yo-yo dieting, or weight cycling. These people seem to be especially vulnerable to metabolic adaptation to a very low-calorie diet.

The importance of maintenance

Weight cycling appears to have other harmful side effects as well, including increased risk of developing hypertension, artery, and gallbladder disease. Research indicates that moderately obese people who don't try to lose weight are healthier than obese people who have experienced frequent fluctuations in body weight.

Some evidence suggests that weight cycling may lead to a "replacement" of muscle with adipose tissue. Each time a person follows a very low-calorie diet, the dieter loses some muscle tissue. Each time the weight is regained, more fat is gained. The person may weigh the same before and after the diet, but be relatively fatter after the diet. It is ironic that dieting behavior may actually be responsible for the obesity the dieter was trying to prevent.

You've heard it before: Successful weight control programs are based on behavior change, on developing nutritious, low-fat, lifelong eating habits you can live with. While some caloric restriction is necessary to create a negative energy balance, a person should not consume fewer than 10 kcals per pound of ideal weight, and never fewer than 1200 kcals. Moderately low-calorie diets do not seem to cause a drop in BMR for most people.

Are the calories burned during exercise worth all of that hard work?

Dieters are fond of reminding us that to burn off the calories in one piece of pecan pie you'd have to run four miles. All that work for 400 measly calories? Forget it. What dieters don't mention is the fact that exercise confers more benefits than simple caloric expenditure. Exercise helps a person build or maintain muscle tissue. Since muscle tissue is metabolically active, the more muscle you have, the higher your BMR. Crash diets that lead to loss of muscle tissue result in a decline in BMR. A moderately low-calorie diet combined with a somewhat vigorous exercise program will result in a negative energy balance and no drop in resting energy expenditure.

Research shows that a person can lose weight on just about any kind of diet. But the variable that best discriminates between people who manage to maintain the weight loss and those who don't is participation in regular exercise. Exercise appears to be vital to maintenance of weight loss.

Exercise also improves mood and self-concept, and decreases one's risk of heart disease, hypertension, and other disorders associated with a sedentary life-style. Aerobic exercise is the best type of exercise for burning calories. Resistance exercise, such as weight lifting, does not burn as many calories as aerobic exercise, but since it increases muscle mass, it can help maintain metabolic rate. People who are very overweight should get their physician's exercise recommendations before embarking on a program of vigorous exercise.

ESSAY 26
RECURRENT UTIs

Urinary tract infections (UTIs) are the most common bacterial infections for women and the second most common illness (after colds) for them. Men get UTIs too, but infrequently. The female's shorter urethra allows bacteria to enter the urinary bladder more easily, where they set up housekeeping and multiply, feeding on the urine stored there. In addition, the urethral and anal openings are closer in women. Eighty-five percent of first-time UTIs are caused by *Escherichia coli* (*E. coli*) bacteria that migrate to the urethra from the anal area. *E. coli* bacteria are necessary for proper digestion, and are welcome in the intestinal tract but cause much pain and suffering if they infect the urinary system.

A number of women develop UTIs several times a month. Antibiotics are very effective in treating most UTIs. The problem comes when the infection recurs very soon after antibiotic treatment has been completed. The patient sometimes feels she is living on antibiotics and can't seem to stay healthy without them. These women must examine their life-styles very closely to track down possible causes for the repeated infections.

The importance of prevention

Personal hygiene is the first line of prevention. Care must be taken to avoid transporting bacteria from the anal area to the urethra. Girls should be taught to wipe from front to back and to wash hands thoroughly after using the toilet. When bathing, women and girls should wash from front to back as well. Mild soaps are often less irritating than soaps with fragrances or deodorants, which may cause allergic reactions. Other perfumed products, such as perfumed toilet paper, feminine deodorant sprays, and even bubble baths can contribute to UTIs. Laundry products such as detergents, bleaches, and fabric softeners leave residues on clothing and may thus cause irritation.

Menstrual blood provides an excellent growth medium for bacteria. Sanitary napkins and tampons should be changed frequently. Some women find that switching from tampons to napkins or from napkins to tampons reduces the frequency of UTIs. During menstruation, a woman's hands come into more frequent contact with the vaginal area as she changes tampons and napkins. Women are used to washing their hands after changing tampons and napkins, but it is just as important to wash them before, especially if tampons are inserted without an applicator. Deodorant tampons and napkins can increase irritation. Superabsorbent tampons absorb the normal secretions of the vaginal wall along with menstrual flow, and women tend to change them less frequently than less-absorbent tampons. They expand to a larger size in the vagina, and may thus be more irritating. Intercourse during days of menstrual flow should be avoided, since blood can get pushed up into the urethra.

People who are prone to UTIs should drink at least 2 to 2.5 liters of fluid daily. It is usually recommended that caffeinated beverages be avoided, since these can be irritating. Cranberry juice increases the acidity of the urine, which may help to decrease bacterial growth, and many women believe drinking cranberry juice helps them get over a UTI more quickly. On the other hand, some people may find that the increased acidity of the urine causes an increase in the burning sensation during urination.

Voiding frequently, every 2 to 3 hours, helps to prevent recurrent UTIs, since it expels bacteria and eliminates the urine needed for their growth. Most people find that it is natural to urinate this often when they drink 2 liters of fluid daily!

One study found that the urinary bladders of some women with recurrent UTIs were stretched from urinating infrequently. A full bladder stretches the bladder wall and compresses the blood vessels in the bladder wall. A decreased blood supply means fewer infection-fighting immune cells. More frequent urination prevents this problem, so women are advised not to "hold it."

Heat and moisture favor bacterial growth. Underwear made of breathable fabrics such as cotton keep the perineum cooler and drier than synthetics. Panty hose and tight jeans can increase heat and humidity.

Honeymoon cystitis

The observation that intercourse is frequently associated with the onset of UTIs led to the term *honeymoon cystitis*. Cystitis is an inflammation of the urinary bladder. Women who find that sex brings on UTIs learn to develop and teach their partners stringent personal hygiene. Both partners should wash hands and genitals before sex and avoid any practices that could introduce bacteria from the anal area into the urethra. Some women find that certain positions are more irritating than others, and partners must communicate with each other to find workable solutions. Women should drink plenty of water before and after sex, and urinate as soon afterwards as possible. This flushes out bacteria that may have entered the urethra.

Diaphragm users may find that the rim of the diaphragm presses on the urethra and prevents complete emptying of the bladder. Since the diaphragm must be worn for several hours after

intercourse, this can prevent bacteria from being flushed out of the urethra and encourage their migration to the bladder. Sometimes switching to a smaller diaphragm or one with a different type of rim solves this problem. The spermicide that must be used with the diaphragm can also be irritating to the urethra. A different method of contraception may need to be adopted.

Estrogens increase the risk of cystitis, since they make the urinary tract more hospitable to bacteria. Hence, UTI risk is somewhat higher for women using the birth control pill, which increases levels of estrogens. Risk also increases during pregnancy. UTI symptoms should be treated aggressively during pregnancy to prevent pyelonephritis, which has been associated with premature birth and low-birth-weight babies.

UTI symptoms may indicate the presence of a sexually transmitted disease such as gonorrhea, chlamydia, or herpes, and these must be ruled out in cases of recurrent UTIs. Sometimes a woman's partner is the source of bacterial transmission, and when UTIs continue to recur, the partner should be tested for *asymptomatic urethritis,* any bacterial infection of the urethra other than gonorrhea. Sometimes treating the partner with antibiotics cures both parties.

ESSAY

27 PROLONGED PHYSICAL ACTIVITY: A CHALLENGE TO FLUID AND ELECTROLYTE BALANCE

Heavy or prolonged physical activity can lead to dehydration and disrupt fluid and electrolyte homeostasis. During physical activity, muscle contraction generates a great deal of heat, up to 100 times more than at rest. The body can get rid of this extra heat by increasing blood flow to the skin, where heat is given off by radiation and convection, and by activating the sweat glands to increase heat loss by evaporation. In very hot weather, radiation and convection do not work, so the body must rely primarily on evaporation to cool itself. Strenuous exercise in hot weather may cause the loss of over 2 liters of water per hour from the skin and lungs. Such losses can lead to dehydration and hyperthermia if fluids are not replaced.

Dehydration

Dehydration refers to a loss of body fluid that amounts to 1% or more of total body weight. It is most common during exercise in the heat, but can also occur during very low levels of physical activity in a hot environment or during strenuous exercise in a thermally neutral environment. Fluid deficits of 5% are common in athletic events such as football, soccer, tennis, and long-distance running. At this level, symptoms include irritability, fatigue, and loss of appetite. Dehydration levels greater than 7% may cause heat exhaustion. Heat exhaustion is characterized by a normal or slightly low body temperature, heavy perspiration, and cool and clammy skin. Muscle cramps, dizziness, vomiting, and fainting may also occur.

With dehydration, water is lost from all body compartments. The decrease in blood volume has deleterious effects on physical performance, since it decreases the amount of blood the heart can pump per beat. Muscles need oxygen to work, so performance decreases as cardiac output is reduced. The body tries to maintain blood volume to the muscles by constricting vessels in the skin, so less heat is lost and body temperature rises. Intracellular electrolyte changes may also occur with dehydration and interfere with optimal performance.

Thirst is the body's signal that its water level is getting too low, and it motivates a person to drink. Unfortunately, thirst is not a reliable indicator of fluid needs. People tend to drink just enough to relieve their parched throats. This is especially true in hot weather, when more fluid than usual is lost through sweating. The thirst mechanism is especially unreliable in children and older adults. Aging also decreases the kidney's ability to retain water when the body needs fluids, which increases the susceptibility of older people to dehydration.

Rehydration

Nutritionists advise that most people can meet their daily fluid needs with 6 to 8 cups of water, juices, and milk. Alcoholic and caffeinated beverages act as diuretics and thus increase fluid needs, so these should not be counted as part of the recommended intake. One of the ways to tell whether or not you are drinking enough fluid is to check the color of your urine. Dilute urine is very pale and indicates sufficient fluid intake.

People experiencing severe dehydration may find that plain water is not the optimal solution. Studies have shown that when a dehydrated person consumes water, the water dilutes the blood as plasma volume is replenished. This removes the feeling of thirst, and protects against low plasma electrolyte levels. In other words, as electrolyte levels drop, the sensation of thirst goes away so the blood will not become any more dilute. The kidney senses the increase in fluids in renal tubules and begins to excrete water. In laboratory studies, subjects consuming plain water after dehydration needed to urinate long before hydration was complete. Although plasma volume increased to some extent when water was consumed, it did not return to its desirable level, and subjects did not rehydrate all body cells and extracellular compartments.

Enter Gatorade. When sodium is taken along with water, dehydrated subjects rehydrate to a greater level than subjects taking only water. Sodium helps to restore plasma volume and to retain water in the blood without inhibiting thirst. This observation has prompted extensive research into the ideal sports drink, designed to help the athlete recover fluid and electrolyte homeostasis.

A typical sports drink contains about 50 to 100 mg of sodium. While some sodium is lost in sweat, the amount is quite small, especially if the person is acclimated to the heat. The National Research Council recommends a daily sodium intake of 1100 to 3300 mg for most adults. Some nutritionists have expressed concern that since most Americans consume 10 to 60 times this daily sodium requirement, sports drinks add insult to injury. Sports drinks certainly aren't necessary for the recreational athlete who plays a leisurely game of tennis doubles or a person who walks briskly for half an hour.

You are a candidate for such a drink if you exercise to the point of dehydration. Your body weight is a good measure of hydration. Weigh yourself before and after exercise. For each pound of body weight lost, drink 2 cups of fluids. Use sports drinks only if you lose more than 1% of your body weight and you are not on a salt-restricted diet.

Sports drinks also contain some amount of carbohydrate. Glycogen is a preferred source of energy during physical activity. Glycogen stores need to be replenished following prolonged exercise, so sports-drinks manufacturers add carbohydrate to some drinks to supply energy. Many of these drinks contain a glucose polymer, an easily digestible form of complex carbohydrate. People performing heavy physical activity for longer than 60 to 90 minutes experience depletion of muscle glycogen. They are able to work better and longer when they consume fluids containing carbohydrate. An athlete should obviously never try something new on the day of a race or contest. Some people have reported gastric distress after consuming sports drinks. Therefore, athletes should use these drinks regularly if dehydration is a problem, and not just on race day.

Carbohydrate supplies calories, and extra calories are made into fat, so exercisers who are watching their calorie intake should not forget to count the calories in these drinks.

Carbohydrate also increases the palatability of the drink. One study showed that when subjects were given unlimited access to either a carbohydrate drink or plain water during recovery from exercise, subjects tended to drink more of the former. If it tastes good, we're likely to drink more of it.

ESSAY

FERTILITY FRAGILITY

28

For most young people today, the biological ability to reproduce comes about 10 years too early. Most prefer to postpone childbearing until their schooling has been completed and they are somewhat settled in their relationships and careers. Since adolescents and young adults generally become sexually active before this occurs, birth control and fear of pregnancy are frequent concerns. Protecting fertility may be the furthest thing from their minds. But eventually, a majority of couples decide to have children. Of these couples, 10 to 15% will have some sort of fertility problem. Many find it ironic that after years of anxiety over late periods, burst condoms, and missed pills, they may discover that becoming pregnant is not so easy after all.

Infertility in females is the inability to conceive; in males, the inability to produce viable sperm in sufficient quantity. About 40% of fertility problems are due to infertility in the male partner; 50% of the time, the female partner is infertile; and in 10% of the cases the problem is shared or the cause is unknown. Problems with fertility have risen dramatically in recent years. Researchers suspect that several factors have contributed to this rise.

Reproductive infections

Infections of the reproductive system often have adverse effects on long-term fertility. These infections are primarily sexually transmitted diseases (STDs) such as gonorrhea, syphilis, chlamydia, and pelvic inflammatory disease (PID). STDs have reached epidemic proportions in North America, especially among adolescents and young adults. A woman's fertility is especially vulnerable, as these infections may cause scarring in the delicate tissue of the Fallopian tubes. Tubes blocked with scar tissue do not allow for the passage of the oocyte or sperm.

Sometimes these infections, especially chlamydia and PID, occur with minor or even no symptoms. By the time they are discovered, damage may already have been done. In order to protect themselves from these infections, women and men are advised to use both condoms and spermicide when having intercourse (unless both are monogamous and uninfected). The condom provides a barrier not only to sperm, but also to bacteria and viruses. The spermicide provides a hostile environment for many pathogens and helps with birth control as well. Men and women should be aware of signs of infection, including exposure to a partner with an STD, pelvic pain, and any unusual discharge. Women should have yearly pelvic exams that include screening for asymptomatic STDs.

Contraceptives

Some women experience a loss of the menstrual cycle after they stop taking oral contraceptives, and have difficulty conceiving. Women who had irregular periods before starting on the Pill are most likely to experience this effect. IUDs slightly increase risk of PID.

Maternal age

It has become increasingly common for women to delay childbearing until they are in their thirties and even early forties. (Men can wait even longer.) Many feel that better financial, marital, and job security will improve their potential to be good providers and parents. Others simply have other things they want to do before having children.

Unfortunately, risk of infertility increases with age, especially in females. Each year that goes by is one more year in which physiological barriers to reproduction can arise. Endometriosis is a good example. This disease involves the abnormal growth of the uterine lining, the endometrium. Endometrial cells may grow on the ovaries, or into the Fallopian tubes, which may become blocked. Almost 20% of infertile women have endometriosis, which typically occurs in women in their thirties and forties who have not had children.

Personal health

Many factors can interfere with the menstrual cycle, especially in younger women. Heavy exercise, low-calorie diets, weight loss, too little body fat, obesity, smoking, alcohol and drug abuse, too much caffeine, and even psychological stress can interrupt ovulation and cause infertility, at least temporarily. Normal cycles often resume once the cause of the problem is discovered and corrected.

A word of caution: An absence of the menstrual bleeding does not necessarily mean a woman is infertile! Many young women erroneously assume they cannot get pregnant since they are not having periods, but it is possible that ovulation is still occurring. Unless the woman is trying to conceive, birth control is essential even in the absence of menstrual periods.

Seeking help for infertility

Couples who have not conceived after a year of trying to get pregnant are generally advised to look for a medical diagnosis. Sometimes couples simply need to keep trying; about half of these couples succeed in conceiving even with no treatment. But since infertility treatment can take several years to produce a successful pregnancy, early investigation is recommended, especially for older couples.

Treatment for infertility begins with medical testing to discover why conception has not occurred. The man's sperm is analyzed, and the woman's menstrual cycle is evaluated to be sure she is ovulating. Hormone levels are checked, and samples of the uterine lining may be taken. The woman's reproductive organs are evaluated with x-rays and laparoscopy. (A laparoscope is a thin telescope to which a laser is attached. It is used to see the reproductive organs and sometimes treat certain problems.)

Medical testing for infertility can be an arduous process, both physically and psychologically. The process includes embarrassing questions, physical discomfort, financial expense, and months of waiting for test results. Sex can become a scheduled and stressful event, rather than an expression of love and desire. Love, understanding, and mutual support are vital for couples undergoing infertility evaluation to prevent the marital discord and decline in sex drive that sometimes occur in this situation.

About half of all couples undertaking treatment for infertility eventually have biological children. In some cases, the physiological barriers to conception may be corrected with medical procedures. When this cannot be accomplished, several alternatives are available, depending on the nature of the infertility. Donor sperm may be used to fertilize the woman's egg if the man is infertile. When for some reason conception cannot occur in the woman's body, various in vitro fertilization (IVF) techniques may help. With IVF, an oocyte is fertilized by sperm in the laboratory. The fertilized ovum is then placed back in the uterus to develop. Oocytes and sperm may come from the biological parents or from donors. If a woman cannot produce oocytes, the couple may opt for a surrogate mother, with the man's sperm fertilizing the egg of another woman, who carries the child and delivers it to the infertile couple after birth. If the woman can produce oocytes but for some reason cannot have a pregnancy, a second woman may provide a "host uterus" for the fertilized egg from the original couple. Many couples become parents by adopting children.

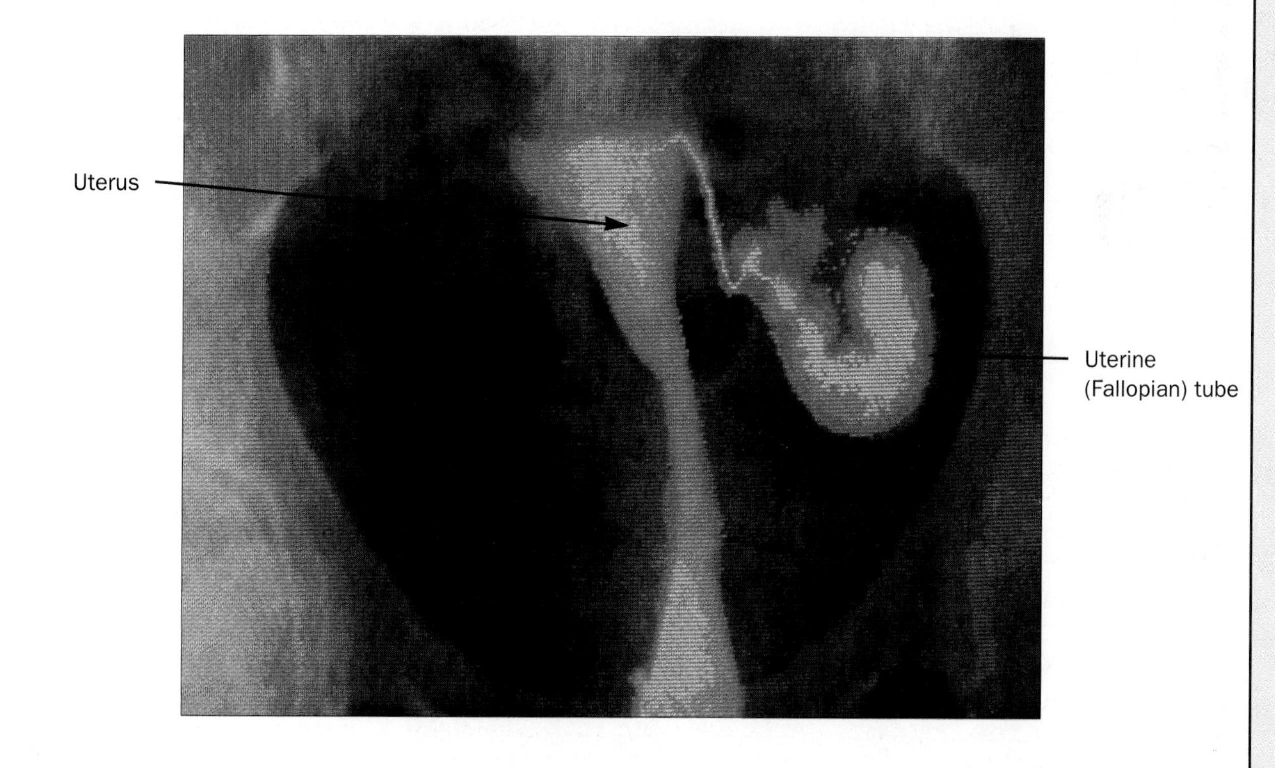

ESSAY

29 EATING FOR TWO

During pregnancy, many of the habits a woman has taken for granted have to be reevaluated as she realizes that she has become responsible for shaping a new life. Some of the most important habits that deserve special consideration are those affecting the nutritional status of the mother-to-be and her developing fetus. Many studies have shown that a woman's diet during pregnancy has important effects on both the mental and physical development of the fetus and her own health as well.

Eating for two means more than taking double portions of dessert. At issue during pregnancy is diet quality, rather than quantity. A pregnant woman is responsible for providing the developing fetus with nutrients, the raw materials necessary for the creation of a new life. Needs for many nutrients increase substantially during pregnancy, while caloric needs increase by only 300 to 500 kcals per day. In other words, there's still not much room for empty calories. Those calories need to be "spent" on food choices that provide the nutrients needed for fetal growth and to support the physical changes that accompany pregnancy.

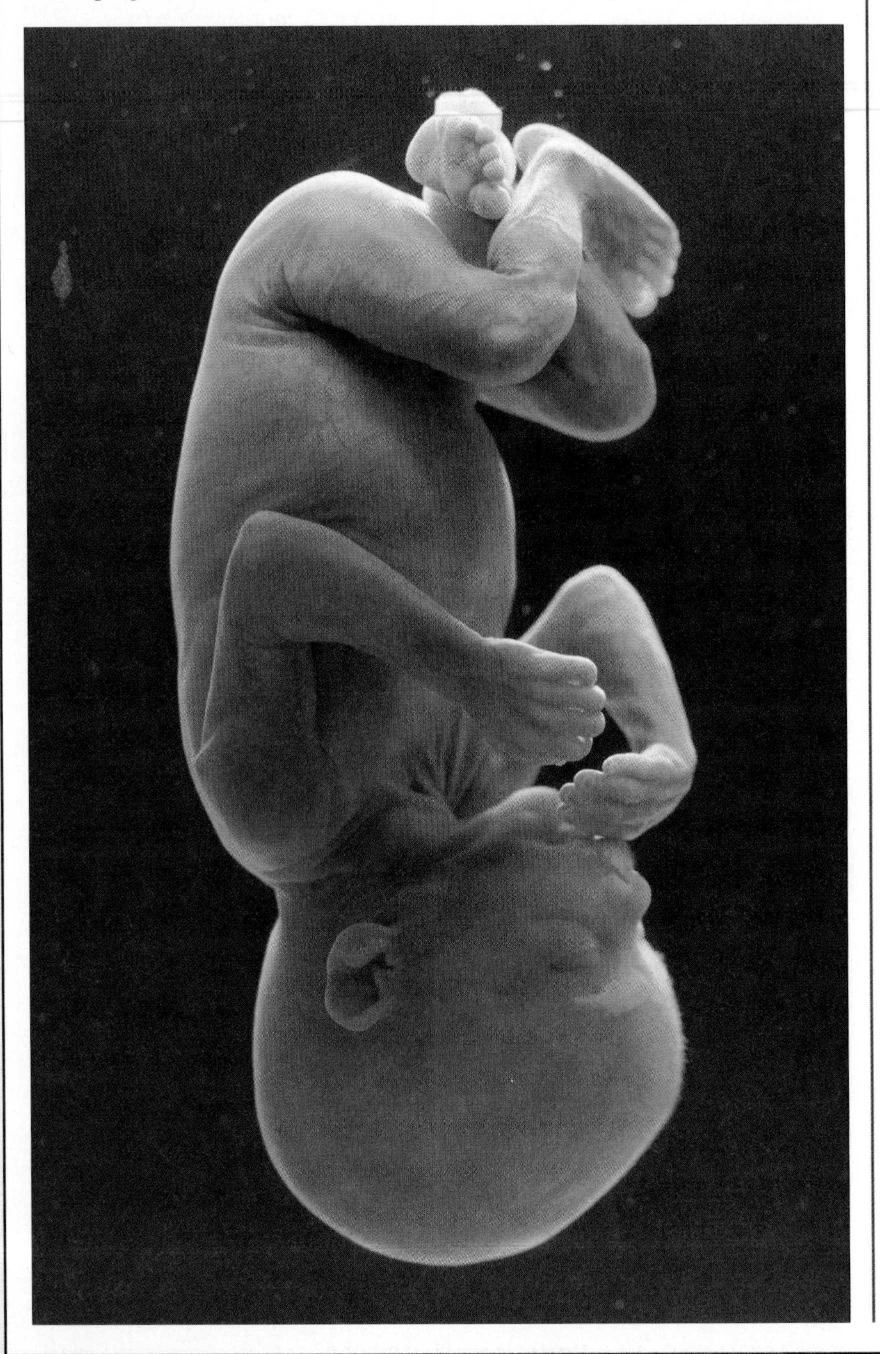

People used to believe that the developing fetus was akin to a parasite that could take whatever nutrients it needed from its host, the mother. We now know this analogy is not accurate. While the placenta is capable of manufacturing some nutrients, such as glycogen and cholesterol, most nutrients are brought to the fetus by the mother's blood. If a nutrient is missing from her bloodstream, fetal development may be compromised. For example, if the amino acids needed for brain development are absent, brain growth is not normal.

Maternal malnourishment is a major cause of low birth weight. Babies who weigh less than 5.5 pounds at birth are 10 times more likely to be mentally impaired than bigger babies and have a much higher risk of illness and death. It's up to the mother to provide the best possible environment for her developing baby. The following are some of the nutrients considered especially vital to the growth of the fetus and the physiological changes occurring in the pregnant woman.

Protein

Amino acids are the stuff of which life is made. Daily protein requirements increase from 50 to 60 g/da during pregnancy. Pregnant women are advised to consume at least three servings of protein foods daily, in addition to their four servings of milk.

Iron

Both the fetus and the mother need iron to make red blood cells. It is almost impossible to obtain the 30 mg of iron per day required during pregnancy from food alone, so iron is a recommended prenatal supplement.

Folic acid

Folic acid is a vitamin required for cell division and development of the fetal nervous system. Research has shown that a woman who receives extra folic acid a month before conception and for the first few months of fetal development has a much lower risk of giving birth to a baby with neural-tube defects such as spina bifida. Folic acid requirements during pregnancy are 400 to 800 µg per day, and because of its importance, folic acid is another recommended prenatal supplement.

Calcium

Calcium is needed for the growth of bones and teeth. Blood calcium level is regulated by a number of hormones, so its level remains fairly constant. If dietary calcium is insufficient, calcium will be released from the mother's bones. That's why a woman is advised to consume four servings of milk and dairy products daily during pregnancy. Women who don't like milk often learn to disguise it in custards and milk-based soups. Teenage mothers, especially those who are still growing themselves, need six milk servings per day.

Fiber

Although fiber is not considered a nutrient, an adequate intake of fiber is important during pregnancy to prevent constipation. Constipation typically increases because of the action of progesterone, which relaxes smooth muscle, and compression of the colon by the growing uterus. Iron supplements, often prescribed during pregnancy, frequently contribute to constipation.

Fluids

As metabolic rate and blood volume increase, so does a woman's need for fluids. Six to eight glasses daily are recommended. Fluid intake should be even greater in hot environments and if the pregnant woman is exercising.

Other considerations

A woman's usual eating practices are often disrupted by pregnancy. Hormonal changes may cause nausea and vomiting ("morning sickness") during the first three months. Since little weight gain occurs during the first trimester, many women don't realize how much the embryo is growing and assume good nutrition is not yet a concern. Even though the embryo is still very tiny, good nutrition is as important as ever. Many women find that many small snacks work better than big meals and manage to get some good nutrition that way.

Toward the end of the pregnancy, the stomach becomes compressed as the uterus pushes up against the diaphragm, and big meals become impossible. Five or six small but nutritious meals usually work best during this time.

Weight gain is often a concern during pregnancy. While medical authorities generally suggest a weight gain of about 25 pounds, weight gain varies a great deal. Recent research looking at weight gain among normal-weight women experiencing healthy pregnancies and giving birth to healthy babies suggests that the upper limits of weight gain may be closer to 35 pounds. In one such group, 75% of the women gained more than 25 pounds. The median weight gain was 33 pounds, and about 25% of the women gained more than 40 pounds with no negative health effects for either the mother or baby. In contrast, women who gain relatively little weight are at risk of delivering low-birth-weight babies.

What happens to the extra weight after delivery? One study found that on the average mothers ended up only 2.2 pounds heavier after the pregnancy. Many obstetricians now feel that the energy spent worrying about weight gain should be replaced with a concern for nutritious food intake.

Page 1 (left): © Jalandoni, Monkmeyer Press Photo.

Page 1 (center): © Bider, PhotoEdit.

Page 1 (right): © Omikron, Photo Researchers.

Page 11: Alper, Stock, Boston.

Page 12: Daemmrich, Stock, Boston.

Page 13: © Hannon, The Picture Cube.

Page 14: Morsch, Stock Market.

Page 17: © Wyman, Phototake.

Page 20: Jean Jackson

Page 25 (top, middle, and bottom): John Moore, © Scott, Foresman.

Page 26: © 1990, Gupton, Stock, Boston.

Page 29: Forestier, Sygma.

Page 30: © Daemmrich, Stock, Boston.

Page 32: © Bider, PhotoEdit.

Page 35: © Rousseau, The Image Works.

Page 36: © Daemmrich, The Image Works.

Page 39: © McIntyre, Photo Researchers.

Page 40: LeDuc, Monkmeyer Press Photo.

Page 43: © Watson, The Image Works.

Page 44: © Jalandoni, Monkmeyer Press Photo.

Page 47: © Freeman, PhotoEdit.

Page 49: Fraser/Victoria Infirmary/SPL, Custom Medical Stock.

Page 50 (left): © Gelderblom, Visuals Unlimited.

Page 50 (right): © Omikron, Photo Researchers.

Page 53 (top and bottom): Gordon T. Hewlett.

Page 54: © 1990, Long, Stock Market.

Page 57: © 1990, Ferguson, PhotoEdit.

Page 59: © Bitters, The Picture Cube.

Page 61: SPL, Photo Researchers.

Page 62: Stevenson, SPL, Photo Researchers.